65.00

# Progress in Rural Policy and Planning

Volume Five
1995

# Progress in Rural Policy and Planning
## General Editor: ANDREW W. GILG

*Regional Editors*:

Robert S. Dilley (Canada) Department of Geography, Lakehead University, Thunder Bay, P7B 5E1, Ontario, Canada

Owen Furuseth (USA) Department of Geography, University of North Carolina, Charlotte 28223, North Carolina, USA

Andrew W. Gilg (UK) Department of Geography, University of Exeter, Exeter, EX4 4RJ, United Kingdom

Philip Lowe and Jonathan Murdoch (Europe) Department of Agricultural Economics and Food Marketing, University of Newcastle upon Tyne, NE1 7RU, United Kingdom

Geoff McDonald (Australasia) Department of Geographical Sciences and Planning, The University of Queensland, Brisbane, Qld 4072, Australia

# Progress in Rural Policy and Planning

Volume Five
1995

*Edited by*
Andrew W. Gilg
and
Robert S. Dilley
Owen Furuseth
Philip Lowe
Geoff McDonald
Jonathan Murdoch

JOHN WILEY & SONS
Chichester • New York • Brisbane • Toronto • Singapore

Published 1995 by John Wiley & Sons Ltd
        Baffins Lane, Chichester,
        West Sussex PO19 1UD, England

        National     01243 779777
        International  (+44) 1243 779777

*Other Wiley Editorial Offices*

John Wiley & Sons, Inc., 605 Third Avenue,
New York, NY 10158-0012, USA

Jacaranda Wiley Ltd, 33 Park Road, Milton,
Queensland 4064, Australia

John Wiley & Sons (Canada) Ltd, 22 Worcester Road,
Rexdale, Ontario M9W 1L1, Canada

John Wiley & Sons (SEA) Pte Ltd, 37 Jalan Pemimpin #05-04,
Block B, Union Industrial Building, Singapore 2057

*British Library Cataloguing in Publication Data*

A catalogue record for this book is available from the British Library

ISBN 0 471 94912 4
ISSN 0956-4187

Typeset in 10/12pt Sabon by Mayhew Typesetting, Rhayader, Powys
Printed and bound in Great Britain by Biddles Ltd, Guildford and King's Lynn

This book is printed on acid-free paper responsibly manufactured from sustainable forestation,
for which at least two trees are planted for each one used for paper production

# Contents

Figures     vii
Tables     viii
Contributors     ix
Abbreviations and acronyms     xi
Preface     xv

**Section I: United States of America**     1
Introduction *Owen J. Furuseth*     3
  1   1993–94, the year in review in US rural planning and policy: of
      mice and men
      *Mark B. Lapping*     5
  2   Sustainability, regional planning and the future of New York's
      Adirondack Park
      *Robert J. Mason*     15

**Section II: United Kingdom**     29
Introduction *Andrew W. Gilg*     31
  3   Annual review of rural planning in the United Kingdom:
      autumn 1993–autumn 1994
      *Andrew W. Gilg*     33
  4   Sustainability, equality, and pluriactivity: the suitability of the
      farm household as a vehicle for rural development
      *Lyneth Davies, Nicholas Mack and Morag Mitchell*     89

**Section III: Europe**     99
Introduction *Philip Lowe and Jonathan Murdoch*     101
  5   European review 1993/4
      *Philip Lowe and Jonathan Murdoch*     103
  6   Towards sustainable tourism for Europe's protected areas —
      policies and practice
      *Rosie Simpson*     125
  7   The changing competitive advantage of rural space
      *Elena Saraceno*     139

**Section IV: Canada**     155
Introduction *Robert S. Dilley*     157
  8   Canada: The rural scene
      *Robert S. Dilley*     159

9  Policy, fisheries management and development in rural
   Newfoundland
   *Rosemary E. Ommer*                                          171

**Section V: Australasia**                                     189
Introduction *Geoff McDonald*                                   191
10  Realigning food power in New Zealand
    *Michael Roche, Richard Le Heron and Anne Pomeroy*         193
11  International protectionism and regional dimensions of changing
    farm financial performance in Australian broadacre cropping
    *Michael Taylor*                                           201

# Figures

2.1 Proposals of the Commission on the Adirondacks in the
Twenty-First Century                                                     17
3.1 UK Environmentally Sensitive Areas 1994                              49
11.1 Agricultural regions defined by the Office of Local
Government and Administrative Services                                   207
11.2 Broadacre cropping: 1984–85 to 1990–91. Change in cash
operating surplus. Beginning trend value                                 208
11.3 Broadacre cropping: 1984–85 to 1990–91. Change in cash
operating surplus. Annual rate of change                                 209
11.4 Broadacre cropping: 1989–90 and 1990–91. Farm financial
performance. Cash operating surplus                                      210
11.5 Broadacre cropping: 1984–85 to 1990–91. Change in rate of
return on labour. Beginning trend value                                  211
11.6 Broadacre cropping: 1984–85 to 1990–91. Change in rate of
return on labour. Annual rate of change                                  212
11.7 Broadacre cropping: 1989–90 and 1990–91. Farm financial
performance. Rate of return on labour                                    213
11.8 Broadacre cropping: 1984–85 to 1990–91. Change in debt/
equity ratio. Beginning trend value                                      214
11.9 Broadacre cropping: 1984–85 to 1990–91. Change in debt/
equity ratio. Annual rate of change                                      215
11.10 Broadacre cropping: 1989–90 and 1990–91. Farm financial
performance. Debt/equity ratio                                           216

# Tables

3.1   Public expenditure on rural planning in the 1990s                37
3.2   Arable area payment scheme 1994–5                                43
4.1   Distribution of pluriactivity among survey households           91
4.2   Pluriactivity in household members by region                    92
4.3   Most common employment sectors for farmers and spouses
      in off-farm work                                                 92
5.1   Population eligible for Objective 5b in the period 1994–9        123
6.1   Key tourism trends                                              127
6.2   Tourism activities generally compatible with protected areas     130
6.3   Tourism activities generally incompatible with protected
      areas                                                           130
6.4   Guidelines for sustainable tourism                              131
9.1   Canadian Atlantic groundfish allocations and catches for the
      years 1978–93                                                   174
9.2   The number of fish processing plants registered with DFO
      for Canada's Atlantic provinces in the years 1981–91            174
10.1  Rural and mixed electorates FPP and MMP                         195
11.1  The changing structure of broadacre farm input costs           204
11.2  Variables used to identify the environmental characteristics
      of the agricultural regions drawn up by the Department of
      Local Government and Administrative Services                    218
11.3  Cash operating surplus, environment and broadacre
      cropping: multiple regression results                           219
11.4  Rate of return on labour, environment and broadacre
      cropping: multiple regression results                           219
11.5  Debt/equity ratio, environment and broadacre cropping:
      multiple regression results                                     220

# Contributors

**Lyneth Davies**, School of Rural Economy and Land Management, Royal Agricultural College, Cirencester, GL7 6JS, UK

**Robert S. Dilley**, Department of Geography, Lakehead University, Thunder Bay, Ontario, P7B 5E1, Canada

**Owen Furuseth**, Department of Geography, University of North Carolina, Charlotte 28223, North Carolina, USA

**Andrew W. Gilg**, Department of Geography, Department of Geography, University of Exeter, Exeter, EX4 4RJ, UK

**Richard Le Heron**, Faculty of Social Sciences, Massey University, Private Bag, Palmerston North, New Zealand

**Mark B. Lapping**, Provost and Vice President for Academic Affairs, University of South Maine, Portland, Maine, 04103, USA

**Philip Lowe**, Department of Agricultural Economics and Food Marketing, University of Newcastle Upon Tyne, NE1 7RU, UK

**Geoff McDonald**, Department of Geographical Sciences and Planning, The University of Queensland, Brisbane, Qld 4072, Australia

**Nicholas Mack**, Department of Agricultural and Food Economics, The Queens University of Belfast, Belfast, BT9 5PX, UK

**Robert J. Mason**, Department of Geography, Temple University, Philadelphia, PA 19122, USA

**Morag Mitchell**, Department of Economics Marketing and Management, Scottish Agricultural College, Aughincruive, Ayr, KA6 5HL, UK

**Jon Murdoch**, Department of Agricultural Economics and Food Marketing, University of Newcastle Upon Tyne, NE1 7RU, UK

**Rosemary E. Ommer**, Institute of Social and Economic Research, Memorial University of Newfoundland, St Johns, Newfoundland, A1B 3X9, Canada

**Ann Pomeroy**, Rural Affairs, Ministry of Agriculture and Fisheries, Wellington, New Zealand

**Michael Roche**, Faculty of Social Sciences, Massey University, Private Bag, Palmerston North, New Zealand

**Elena Saraceno,** Centro Ricerche economic-socialie, riva Bartolini 18, 33100 Udine, Italy

**Rosie Simpson,** Countryside Commission, Crescent Place, Cheltenham, GL50 3RA, UK

**Michael Taylor,** Department of Geography, University of Portsmouth, PO1 3HE, UK

# Abbreviations and acronyms

## Section I: USA

| | |
|---|---|
| APA | Adirondack Park Agency |
| ATSDR | Agency for Toxic Substance Disease Registry |
| EPA | Environmental Protection Agency |
| FFA | Farms for the Future Act |
| GATT | General Agreement on Trade and Tariffs |
| LULU | Locally Unwanted Land Uses |
| NAFTA | North American Free Trade Agreement |
| NIMBY | Not in MY Backyard |
| USDA | United States Department of Agriculture |

## Section II: UK

| | |
|---|---|
| ADAS | Agricultural Development and Advisory Service |
| ADC | Association of District Councils |
| AONB | Area of Outstanding Natural Beauty |
| ATB | Agricultural Training Board |
| CADW | Welsh Historical Monuments |
| CAP | Common Agricultural Policy |
| CCP | Countryside Commission Publication |
| CLA | Country Landowners Association |
| Cm | Command Paper |
| CPRE | Council for the Protection of Rural England |
| DAFS | Department of Agriculture and Fisheries for Scotland |
| DOE | Department of the Environment |
| EAGGF | European Agricultural Guarantee and Guidance Fund |
| EC | European Community |
| ECU | European Currency Unit |
| EIA | Environmental Impact Assessment |
| ERM | Exchange Rate Mechanism |
| ESA | Environmentally Sensitive Area |
| ESRC | Economic and Social Research Council |
| ESU | European Farm Business Size Unit |
| EU | European Union |
| FEOGA | French version of EAGGF |
| FOE | Friends of the Earth |
| GATT | General Agreement on Tarriffs and Trade |
| GDO | General Development Order |

| | |
|---|---|
| GDP | Gross Domestic Product |
| HC | House of Commons (Paper) |
| HL | House of Lords |
| HLCA | Hill Livestock Compensatory Allowance |
| HMSO | Her Majestys Stationery Office |
| IACS | Integrated Administration and Control System |
| JAEP | Joint Agriculture Environment Programme |
| LEADER | Links between Activity for the Development of the Rural Economy |
| LINK | Technologies for Sustainable Farming Systems |
| LFA | Less Favoured Areas |
| MAFF | Ministry of Agriculture, Fisheries and Food |
| MMB | Milk Marketing Board |
| MP | Member of Parliament |
| MPG | Minerals Policy Guidance |
| NCC | Nature Conservancy Council |
| NFU | National Farmers Union |
| NNR | National Nature Reserve |
| NPPG | National Planning Policy Guidance |
| NRA | National Rivers Authority |
| NSA | Nitrate Sensitive Areas |
| NZV | Nitrate Vulnerable Zones |
| OECD | Organisation for Economic Cooperation and Development |
| PAN | Planning Advice Note |
| PPG | Planning Policy Guidance |
| RDC | Rural Development Commission |
| RPG | Regional Planning Guidance |
| RSPB | Royal Society for the Protection of Birds |
| RSPCA | Royal Society for the Prevention of Cruelty to Animals |
| RTPI | Royal Town Planning Institute |
| SAC | Special Areas of Conservation |
| SI | Statutory Instrument |
| SSSI | Site of Special Scientific Interest |
| SPA | Special Protection Area |
| TCPA | Town and Country Planning Association |
| UCL | University College London |
| UK | United Kingdom |
| VAT | Value Added Tax |
| WWFN | World Wide Fund for Nature |

## Section III: Europe

| | |
|---|---|
| AMS | Aggregate Measure of Support |
| BSE | Bovine Spongiform Encephalopathy |

CADISPA       Conservation and Development in Sparsely Populated Areas
CAP           Common Agricultural Policy
DG X1         Environment Directorate of the European Commission
DGXV1         Regional Policy Directorate of the European Commission
EC            European Community
EEA           European Environment Agency
EFTA          European Free Trade Area
EKHF          Environmental Know-How Fund
ERM           European Exchange Rate Mechanism
ETB           English Tourist Board
EU            European Union
FNNPE         Federation of Nature and National Parks of Europe
GATT          General Agreement on Tariffs and Trade
GDP           Gross Domestic Product
IUCN          International Union for the Conservation of Nature
LIFE          Financial Instrument for the Environment
NIC           Newly Industrialised Countries
OECD          Organisation for Economic Cooperation and Development
RDC           Rural Development Commission
UK            United Kingdom
WWF           World Wildlife Fund

## Section IV: Canada

ALR           Agricultural Land Reserve
CORE          Commission on Resources and Environment
DFO           Department of Fisheries and Oceans
EA            Environmental Assessment
ERCB          Energy Resources Conservation Board
ESC           Ecological Science Centre
IRP           Integrated Resources Plan
ITQ           Individual Transferable Quota
MNR           Ministry of Natural Resources
OMB           Ontario Municipal Board
PALC          Provincial Agricultural Land Commission
PIRPAP        Progress in Rural Policy and Planning
PQ            Parti Québecois
SOE           State of the Environment
UI            Unemployment Insurance

## Section V: Australasia

ABARE         Australian Bureau of Agriculture and Resource Economics
ACPC          Agri-Commodity Production Chains

| | |
|---|---|
| ADC | Agriculture Development Conference |
| APMB | Apple and Pear Marketing Board |
| ARIS | Australian Resource Information System |
| CAP | Common Agricultural Policy |
| CSIRO | Commonwealth Scientific and Industrial Research |
| EEP | Export Enhancement Programme |
| EU | European Union |
| FPP | First Past the Post |
| MAF | Ministry of Agriculture and Fisheries |
| MMP | Mixed Member Proportional |
| MP | Member of Parliament |
| PM | Prime Minister |
| WFF | Watties Frozen Foods |

# Preface

*Progress in Rural Policy and Planning* is now an established periodical which is published on an annual basis. Although PIRPAP, as it is known to its editors, has now created its own identity and reputation, rural planning aficionados will recognize that three of the editors worked on the former *International Yearbook of Rural Planning*, and one of the other two editors was a contributor. Going further back, Andrew W. Gilg, the general editor of PIRPAP, was, of course, the editor of the *Countryside Planning Yearbook* between 1980 and 1986, when it expanded from its UK base into the *International Yearbook*.

There is therefore some continuity between PIRPAP and its predecessors, notably in the UK section. This partly reflects the nature of the UK with its national legislature, which allows an attempt to be made to produce a comprehensive review. In the other sections, however, we are dealing either with federal systems as in Australia, Canada and the USA or with over 20 countries as in the case of Europe. The editors in these countries have therefore come up with a more flexible structure for their sections than the UK, which follows a slimmed-down yearbook format.

The aims for PIRPAP are to provide a regular review of both the progress of rural policy development and the implementation of these policies on the ground. These reviews can take three broad forms: first, considered reviews of developments over a fairly long time-span, normally at least 10 years; second, reviews of events over a much shorter time-span, normally two years or less; and third, the core annual review for each geographical section. PIRPAP is not meant to be a primary research journal, but rather an intermediate stage between such a journal and an abstract or yearbook type of publication. A prime aim is to provide a framework of information for those people working in one branch of rural planning who need access to information in other areas but do not have the time to research new ideas for themselves. In the longer term the aim is to provide a comprehensive record of the development of rural planning as an integrating discipline. In choosing material and authors the emphasis is thus on recording 'progress' in the discipline.

Within this emphasis on 'progress' the contents of PIRPAP embrace all the issues involved in the eclectically broad area of rural policy and planning. Rural policy is defined as legislation, government advice, government or judicial decisions involving case law or ministerial precedent, statements of policy by government organizations and the arguments of rural policy pressure groups. Rural planning is defined as the implementation of these policies on the ground via plans, land-use designations, controls, quotas, grant aid, loans, incentive schemes, capital provision and programmes of various kinds. Clearly both these activities overlap and the definitions are provided only as a guide and not as a complete or rigid list. The editor is working on a model of rural policy and planning and hopefully this will be used as a guide for future contributors.

In a similar vein, the coverage of issues and topics is loosely defined to cover the planning of extensive land uses and social economic provision in rural (non-urban low population density) areas on the following topics: agriculture; forestry; conservation; recreation and tourism; extractive industries; employment; transport; general land-use planning; and social issues. The areal coverage is confined to five regions of the world purely because they have a common rural structure and a set of rural planning systems which are broadly the same, thus allowing comparative work to be assessed and a broad sense of unity for papers from otherwise quite spatially far-apart parts of the world. However, now I have edited five volumes of PIRPAP I have been struck by the diversity within similarity that is found in the nearly 100 chapters contained in these five volumes. In each region of the world a considerable jargon has developed which is not easily understood by the other regions. In other words, each region has a broad suite of powers and organizations that are similar but yet very diverse in their own detailed laws and interpretation. We have much to learn from each other, and here the Maastricht agreement which takes the European Community forward into the European Union has a message, namely, broad structural powers for the whole of the European Union, but devolution of day-to-day, and bread-and-butter issues to local areas, under the principle of subsidiarity.

In a reversal of normal alphabetical order, material from the five regions of the world included is presented in reverse order, namely: the USA; UK; Europe; Canada; and Australasia. The material is presented by region since it has been edited this way, rather than thematically. Comments and offers of papers can therefore be most effectively sent to the regional editors, rather than to me as general editor. In this role it remains only for me to thank all those people who have worked so hard to bring the first four issues of PIRPAP to fruition. It is now up to you, the reader, to tell us how we can further improve PIRPAP and hopefully contribute to its continued development. If you haven't already done so complete the questionnaire contained at the end of Volume Four or write to me in Exeter.

Andrew W. Gilg
Exeter
16 December 1994

# Section I:
# United States of America

edited by
*Owen J. Furuseth*

# Introduction
## *Owen J. Furuseth*

The events since the last issue of *Progress in Rural Policy and Planning* (PIRPAP) can only be described as topsy-turvey. During most of 1994, US rural policies at all levels seemed headed at various speeds towards increasingly 'green' futures. From the top down, governmental and private sector organizations and private sector organizations operating across the agricultural, land use and natural resource spectrum were increasingly focusing their energies on 'environmentally friendly' and sustainable strategies. For example, the preliminary policy directives and program guides for the forthcoming 1995 Farm Bill, discussed by Darrell Napton in the last issue of PIRPAP, are powerful evidence of this trend.

The political events of the last two months of 1994, however, raise the spectre of a complete reversal in policy direction. Riding a tide of voter anger and disgust with politics as usual and the increasing power of the centre (Federal Government) over individual rights, the Republican Party swept the November elections at the Federal level and gained governing status in nearly half of the state legislatures.

While President Clinton and his party remain in control of the executive institutions, the Republican Party now controls the US Senate and House of Representatives (the legislative branch) for the first time in 40 years. In particular, many of the leaders of the new Republican Congress are promising a fundamental restructuring of federal policy making, marked by less federal government involvement in local government and business decision making.

Although rural land and resource management, specifically, and environmental policy initiatives, in general, were not widely debated issues, the Republican political juggernaut was partially fuelled with the votes and finances of an increasingly active anti-environmental movement. Operating under the slogan of the need for environmental balance and the wise use of resources, a coalition of timber, minerals, ranching, and land development interests have been an active counter-point to the recent greening of rural policy. Mark Lapping has been following the efforts of the wise use movement in his previous reports for previous issues of PIRPAP. For instance, his analysis in Volume Three (1993) has proved prophetic.

*Progress in Rural Policy and Planning*, Volume Five. Edited by Andrew W. Gilg
© 1995 Editors and contributors. Published 1995 by John Wiley & Sons Ltd.

In the wake of the Republican congressional control, the future direction of rural land use and resource policy is clearly in doubt. Work on the 1995 Farm Bill, renewal of the Endangered Species Act and other federal policies will be delayed and revised. Just how far and how much of the wise use movement ideology will penetrate the policy-making process is unknown as the year ends. Senator Phil Gramm of Texas, a Republican leader, and a supporter of wise use principles (as well as an announced candidate for President in 1996) has declared that sustainable development is dead, and that individual property rights will be reinstated as the dominant principle for resource management. The roller-coaster ride metaphor that has characterized US natural resource and environmental policy direction over the past 30 years continues.

The US section in this year's PIRPAP is divided into two chapters. In the first chapter, Mark Lapping presents his usual keen review of the American rural policy scene. This year's report includes the latest actions on the statewide front as well as federal initiatives, and several new, critical judicial decisions. Perhaps the most followed rural policy issue over the past year was the Disney Company's proposed historical theme park slated for northern Virginia. As Lapping notes, 'the mouse that roared' saga was prototypic of a growing locally based rural assertiveness in economic development decision making. The current changes in political climate and their results should provide Lapping with ample material for Volume 6. We look forward to his assessment of the impacts of recent changes on rural policy.

The second chapter by Robert Mason examines the establishment and policy evolution of New York's Adirondack Park. The Adirondack Park is unique on the American scene as a 'greenline' regional park, but the development of the park has been fraught with controversy and policy disarray since its establishment. A long-time observer of the Adirondack experience, Mason provides an insightful analysis of the circumstances and events that have affected the park. Ironically, many of the points he raises in his chapter find larger application in the current debate between wise users and sustainability advocates.

# 1 1993–4, the year in review in US rural planning and policy: of mice and men

*Mark B. Lapping*

## Federal issues

The past year in rural planning and development policy, at least from the federal perspective, was largely uneventful. This is the result of two factors: first, the concentration of the Clinton Administration on several key national policy agenda items, such as health care reform, the North American Free Trade Agreement (NAFTA) and the GATT negotiations, and the federal budget; and second, ongoing discussions on the upcoming 1995 Farm Bill which will likely reshape agricultural and rural policy for the remainder of the 1990s and beyond. While each of these has genuine ramifications for the future of rural America, the inability of the Administration to fully reach closure on most of these matters leaves many things simply hanging in something approaching suspended animation.

In last year's annual review mention was made of the near-heroic efforts of the Secretary of the Interior, Bruce Babbitt, to craft some degree of consensus on the future of federal land policy. The Interior Department manages huge portions of the country, especially in the West, and includes such agencies as the Fish & Wildlife Service and the Bureau of Land Management, among others. The issues of grazing fees for ranchers and other fees and taxes for miners, both of whom utilize public lands and too rarely remediate the sites they control, have been at the forefront of the overall federal review of public lands policy. (for European readers, 'remediate' is an American term for 'repair' or in this instance 'to correct the environmental degradation of the sites'.)

Under Babbitt's leadership a number of regional panels will be created to advise the federal government on specific policies for specific large pieces of land. Ranching interests have been particularly strong in their opposition to any outright ban on the use of public lands for cattle grazing. Likewise, they have opposed proposed increases in fees charged for grazing rights. Because

*Progress in Rural Policy and Planning*, Volume Five. Edited by Andrew W. Gilg

the industry is a significant one in many states, and it is well organized and well funded, the Interior Department has come under mounting pressure to abandon its initiatives. Some environmental groups, to the contrary, have been pushing Babbitt in the other direction, a total ban on grazing in some specific places. By mid-February the Secretary was forced to declare that 'I really don't want people on these regional advisory councils who are dedicated to the notion that we want the range cattle-free . . . They are outside the parameters' (Cushman, 1994). The matter is still largely unresolved and has been left for resolution until after the mid-term national elections which are approaching.

Attempts to control the pollution and the loss of the Florida Everglades, which led to a near-historic agreement last year sponsored by the Clinton Administration, culminated in the passage of state legislation, the Everglades Forever Act (Florida Statute 373.4392), in the early spring (Galagher, 1994). Though falling far short of what was initially proposed by the federal government and environmental groups, Florida Governor Lawton Childes was able to fashion a compromise which will force a number of changes on the state's sugar cane and vegetables industries, the primary polluters and developers in the region. Called by some a 'giveaway' to industry, and by others an 'environmental black hole' the agreement nevertheless constitutes the very first concerted effort to protect this valuable wetlands resources.

The reauthorization of the federal Clean Water Act (Public Law 92-500) has also required that serious compromises be made. Yet the emphasis within the discussion on the growing necessity for watershed planning may be one of the more important and enduring initiatives of the year. With its roots in the ideas of the Regional Planning Association of America in the early part of this century, and the experience of the Tennessee Valley Authority since the 1930s, watershed management can both transcend political boundaries and promote the analysis of cumulative impacts of incremental land use changes upon water quality in an entire river basin. Though the details of this approach must still be worked out, it is worth noting that with this initiative to give the federal Environmental Protection Agency (EPA) a greater role in bringing state and local governments into compliance with established rural water quality and quantity goals, the national government has only now come to a place where a number of states have been for some time (Goldfarb, 1994).

Elsewhere on the national level the ongoing reorganization of the US Department of Agriculture continues. This will reduce the number of local USDA offices in the countryside but will promote greater efficiencies, so it is argued, by making the remaining offices 'one-stop-shop' for the administration of most federal agriculture programs and grants. The overall budget of the USDA will be reduced significantly by these and other reorganization measures.

## Agricultural land loss

The Economic Research Service of the USDA released yet another report which declared that nationally the threat of agricultural land loss due to urbanization was not a significant issue or problem (Vesterby *et al.*, 1993). This study generated significant debate and considerable disagreement from those who are concerned about the issue of farmland loss in local areas and others who see distinctive local and regional food resources in jeopardy (Bowers, 1994a). The extent of the loss of farmland has been a matter that has been hotly debated since the early 1980s when the National Agricultural Lands Study (1981) published its pioneering studies which documented the degree of farmland loss. Since that time most federal specialists have argued that nationally the amount of farmland loss is insignificant while those at the local level usually take the opposite view. Part of the problem lies in the natural tendency of federal agencies to agglomerate statistics and to look 'at the whole picture'. In this way lands are frequently seen as little more than interchangeable production factors. The loss of prime agricultural land in one place can theoretically be balanced by new planting or grazing elsewhere or other rises in productivity. But the real impacts of farmland loss are realized locally, both within and without the agricultural community, and national and even regional overviews very often mask these important trends and concerns. It is for this reason, in part, why discussions on the statistical extent of land loss to urbanization and other factors remains a matter of heated debate.

## Farmland preservation

As is usually the case in the United States, most of the important events in the area of rural land use and planning occurred at the local and state levels. With the absence of strong federal involvement the states are clearly defining where the nation is going in terms of the protection of agricultural land. In those states witnessing strong urbanization pressures, namely those of southern New England, the mid-Atlantic states, the upper Midwest and the Pacific Northwest, preservation efforts through land acquisition and easement purchases are common. While funding has been restrained the editor of the *Farmland Preservation Report* has nevertheless established that over 700 000 acres of highly vulnerable farmland have been permanently protected through such measures (Bowers, 1994b). Largely funded by government bonds, grants, and taxes assessed on land transfers, these programs — variously described as purchase of development rights or conservation/agricultural easement acquisitions — have permitted planners to acquire the development rights on a considerable number of strategically important farms. By 'retiring' these farms from the development market, a spatial curb on development

pressures has been created. Further, while individual farms have been purchased everywhere, one of the goals has been to preserve groups of farms
thus providing the 'critical mass' and economies of scale necessary to support
local agribusinesses. These, in turn, help to secure the continued economic
vitality of agriculture in a given region or area. Of the states involved in such
programming only Vermont has benefitted from direct federal aid. Under the
provision of the 1990 'Farms for the Future Act' (FFA) the federal government provides loan assistance for farmland preservation. Vermont is the only
state able to leverage federal resources under this law since it is the sole pilot
jurisdiction under the program. The FFA creates a partnership between the
USDA's Farmers Home Administration, local financial institutions and banks
in Vermont, and the State of Vermont. Funding to support the acquisition of
easement purchases has been increased in this manner through the pooling of
resources, risks to individuals have been reduced, and a partnership to
promote farming has been nurtured. According to the program's local
director, James M. Libby, the FFA in Vermont is both workable and efficient
and has promoted greater local interest and commitment to farmland
preservation (Bowers, 1994c). Farmland preservation organizations and
agencies throughout the country have made the extension of the FFA to other
states a top priority.

In the upper Midwestern states of Wisconsin, Michigan, and Minnesota,
farmland preservation efforts have been geared to the extension of tax relief
to farmers along with land-use restrictions (Freedgood, 1994). Typical of the
ubiquitous differential taxation schemes to help farmers reduce property tax
liability, these programs differ somewhat in tying the extension of tax benefits
to commitments on the part of farmers to retain land in active production for
set periods of time. In this way these states continue to provide farmers with
direct financial relief and to secure land in farming uses. Only Wisconsin has
also wedded enrollment in this program to active farmer participation in
agricultural zoning programs which promise even more long-term protection
against urbanization and land conversion pressures. To date, 43 of the state's
72 counties have adopted exclusive agricultural zoning ordinances to comply
with the terms of the program. However nearly all the counties implementing
such exclusive agricultural zoning schemes are predominantly rural in nature.
Analysts have found that the effectiveness of the program to limit land
conversion on the urban fringe is limited at best (Bowers, 1994d).

Oregon's land-use policy has long been considered a pioneering effort.
Utilizing exclusive farm-use zones, urban growth boundaries, and other
measures, Oregon's program has been singularly innovative and effective. As
Mitch Roshe of the Oregon Land Conservation and Development Department has noted, 'It's very clear that urban growth boundaries have been
effective. You can't build a shopping center on farmland in Oregon — it's
illegal. The harshest critics would say the program has had a significant
impact on protecting farmland' (Bowers, 1994e). Yet even in Oregon efforts

are being made to erode the effectiveness of the program by loosening definitions of permissible uses in exclusive agricultural use zones. Opponents of the Oregon program hope that by making incremental changes in the program over time its effectiveness in limiting development in certain areas will be reduced, thus freeing up more land for housing and commercial growth.

In California, the nation's foremost agricultural state, protection of farmland has been the responsibility of the Department of Conservation through the California Land Conservation Act, commonly known as the Williamson Act (California Government Code Section 1200, 1965). By combining tax relief with some restrictions upon land-use conversion, the Williamson Act has been subject to numerous assessments and evaluations. Most serious students of the program recognize that it has largely failed to protect lands on the urban fringe or in the Silicon Valley south of San Francisco (Bowers, 1994f). The agriculturally significant Central Valley has witnessed a loss of well over a half million acres since the 1980s. This has led the American Farmland Trust to declare the Central Valley one of the nation's most threatened agricultural regions. The Trust also included California's coastal farming regions in this category (Lapping, 1994). The present conservative leadership of the state does not see the protection of California's agricultural land resources as a major policy issue. Thus, the momentum has passed to the localities and counties. Leading the way is Marin County, situated just north of San Francisco. Here the Marin Agricultural Land Trust has managed to permanently preserve more agricultural acreage than any other county in the nation. Sonoma County, which includes part of the state's world-class wine growing region, has instituted an agricultural preservation district which is supported by a one-quarter cent sales tax levied on all purchases within Sonoma County. Elsewhere in California Stanislaus, San Luis Obispo and Sacramento Counties have each adopted 'agricultural buffer zone' ordinances which separate working farmlands from other types of land uses. These vary from 500 feet to 100 feet in width but the goal remains the same: to provide some physical edge which delimits the extension of intensive land uses in farming areas (Bowers, 1993).

## Agricultural zoning

'Right to Farm' and agricultural zoning laws continued to be challenged in the courts. In the case of *Finlay v. Finlay* (856 P.2d 183, Kansas, 1993) the Kansas Court of Appeals ruled that the state's right-to-farm law did not protect a farmer against a nuisance suit brought by a neighbor whose residential use of the land predated the farmer's nuisance generating activity (Crowley, 1993a). Though the farmer's land was used as a farm prior to the creation of an adjoining residential property, the farmer subsequently changed his farm from a general-purpose enterprise to a small animal feedlot operation. This activity, his neighbors successfully argued, created great problems for them. Because the residential use existed prior to the establishment of the feedlot operation, the right-to-farm law could not offer the farmer protection. In Massachusetts the state's right-to-farm law was invoked to protect dog breeding as an agricultural use in a ruling in the state's Court of Appeals (*Town of Sturbridge v. McDowell* (624 N.E. 2d 114, Massachusetts, 1994)). While it was argued that the law was not intended to protect such an activity, and that it was not a standard agricultural practice and use, an expansive definition of 'agriculture' was applied by the court when it allowed dog breeding to continue (Crowley, 1994a). In Illinois, in the case of *Harvard State Bank v. County of McHenry* (620 N.E. 2d 1360, Illinois, 1993), the Appellate Court upheld the right of a county to utilize agricultural zoning to protect farmland from being changed into a housing development. Importantly, the court found that while it was true that the economic value of the land in question would be greater in a more intensive use than agriculture, the diminution in value did not constitute a 'taking' nor was it an adequate reason for nullifying the agricultural zoning ordinance (Crowley, 1993b).

## Northern Forest

Efforts to protect the 26-million acre 'Northern Forest', which stretches across the Adirondacks of New York through the great wilderness lands of Maine, culminated in the completion of the studies and public hearings of the Northern Forest Task Force, funded by the federal government and the States of New York, Vermont, New Hampshire, and Mine. This regional resource consists of unique and remarkable wilderness areas, national forest lands, and large expanses of heavily logged private forest lands. However, poor forest management practices, land development, habitat destruction and water pollution in the Northern Forest have been the focus of public concern for a number of years. Conditions in this vast region have been documented by Mitch Lansky in his *Beyond the Beauty Strip* (1992), and in numerous studies by environmental organizations and groups. Described by one major participant, the noted forest policy expert Carl Reidel of the University of Vermont, as 'a dizzying array of schemes intended to find some common ground between calls for outright public acquisition and industry's fears of regulation', the work of the Task Force ended with few substantial recommendations and essentially returned the problem to the federal government and the states. Reidel also termed the exercise as one akin to 'rearranging the deck chairs on the *Titanic*' (Williams, 1994). Many sense that time is running out for this vast rural land and water resource. The issues were well defined by Mollie Beattie, a native of the region who now heads the US Interior Department's Fish and Wildlife Service, when she concluded:

Pretty soon the time for comfortable data-gathering and mapping will be over. There will be only two choices. The first will be to buy up a couple of reserves and offer up a few standard economic-development programs; effectively to declare victory and get out. The second will be to take a risk that is as big as the Northern Forest itself, that of truly conserving all of this special place before it is gone, without our old techniques, out old suspicions, or our territorial myopia. No more metaphors and maps. Think big. Trust each other. Do something bold (Williams, 1994).

## Environmental justice

The 'environmental justice' movement is an emerging one which seeks to redirect the location of environmentally questionable and dangerous facilities and land uses away from low-income and minority communities, both rural and urban. A series of federal studies published during the year have confirmed what many community activists have long argued, namely that toxic production and waste-disposal sites tend to be concentrated in poor, minority communities in an overwhelming number of cases. The most important of these studies were executed by the federal Environmental Protection Agency (EPA) and the Agency for Toxic Substances Disease Registry (ATSDR) in the Chattanooga, Tennessee, region. These assessments demonstrated strong statistical correlations between diseases and pariah land uses as reflected in racial and income variations (Crowley, 1993). Studies such as these, together with numerous calls for federal action, led President Clinton to issue an Executive Order (12898) requiring all federal agencies to develop a strategy to address these environmental justice issues through a coordinated agency task force (Crowley, 1994). This can be seen as a first important step in a process which will invariably deal with such issues as LULUs (Locally Unwanted Land Uses), the NIMBY syndrome (Not In My Back Yard), and the very nature of the distributional and equity aspects of land-use decision making.

## Disney history park

Finally, one of the more curious countryside planning issues of the year was occasioned by the Walt Disney Company's proposal to create a new American history theme park in a rural area not far from Washington, DC. In close proximity to some of the major battlefields of the Civil War, the Disney proposal to build a huge 1000-acre historical theme park with hotel and convention center facilities, and at a cost of over $650 million, met with fierce local opposition. The 'Take a Second Look' campaign against the proposal was led by the Piedmont Environmental Council (a long-active group in farmland preservation), a number of state legislators, many local residents, some prominent historians, and even members of Congress who saw in the development both a threat to the environmental integrity of the area and an assault on national battlefields and historic memory. Nonetheless, state officials, led by recently elected Governor George Allen, Jr, supported the proposal as an economic development boon and succeeded in passing a special appropriation of over $40 million for new highway, water and sewage, and other infrastructure developments targeted for the Haymarket area to support the Disney project (Crowley, 1994c). As part of the price of gaining support for the appropriation the Governor promised a full environmental review of the highway impacts of the project, estimated to take 18 months to complete. In local hearings to evaluate the proposal it rapidly became clear that Disney's projections of local traffic impacts, for example, were seriously underestimated. This caused consternation over the entirety of Disney's presentation (Crowley, 1994d). Though later withdrawn, a lawsuit was filed in the federal court to block the development, and even the National Trust for Historic Preservation got into the fray by rebuking in public the head of the Disney Company for seeking to destroy historically important landscapes by promoting urban sprawl (Crowley, 1994e). Throughout the summer of 1994 the 'Mouse Wars' continued with charges levelled at both local officials and Disney executives concerning alleged failures to disclose inside deals, cover-ups of the real costs of the project to the communities and county, and outright political pay-offs. But claiming that they were not bending to the opposition, in late September Disney announced that it was abandoning its Prince Edward County site and that it would look for another one in Virginia upon which to build a much-scaled-down version of their initial proposal (Associated Press, 1994).

Perhaps the failure of the Disney Company to build its proposed development provides us with a fitting metaphor for the year in review: the Mouse may have roared but in the end it, too, lost nerve and not very much happened or changed!

# References

Associated Press, 1994, 'Disney to seek new site for proposed theme park', *Portland (Me.) Press Herald*, 29 September: 8A

Bowers, D., 1993, 'More localities enacting agricultural buffer ordinances', *Farmland Preservation Report*, 4(1): 1 and 6

Bowers, D., 1994a, 'USDA/ERS Study: farmland loss is not a national concern', *Farmland Preservation Report*, 4(8): 3, 6–7

Bowers, D., 1994b, 'Nation's preserved farmland: 730,213 acres', *Farmland Preservation Report*, 4(6): 1

Bowers, D., 1994c, 'Vermont: farms for the future program', *Farmland Preservation Report*, 4(9): 1 and 6

Bowers, D., 1994d, 'Wisconsin', *Farmland Preservation Report*, 4(5): 1–2

Bowers, D., 1994e, 'Oregon: still defining farmland protection', *Farmland Preservation Report*, 4(5): 7

Bowers, D., 1994f, 'California', *Farmland Preservation Report*, 4(5): 7

Crowley, M., 1993a, 'Finley v Finley', *Land Use Law Report*, 21(21): 166

Crowley, M., 1993b, 'Harvard State Bank v County of McHenry', *Land Use Law Report*, 21(25): 200

Crowley, M., 1993c, 'Chattanooga studies raise environmental justice issues', *Land Use Law Report*, 21(25): 195

Crowley, M., 1994a, 'Town of Sturbridge v McDowell', *Land Use Law Report*, 22(3): 35

Crowley, M., 1994b, 'Clinton signs environmental justice executive order', *Land Use Law Report*, 22(4): 28

Crowley, M., 1994c, 'Virginia legislators eye theme park costs warily', *Land Use Law Report*, 22(2): 10

Crowley, M., 1994d, 'Theme park's zoning application shows tripled traffic flows', *Land Use Law Report*, 22(1): 5

Crowley, M., 1994e, 'The drumbeat of opposition', *Land Use Report*, 22(9): 71

Cushman, J.J., 1994, 'A consensus approach on land use: Babbitt tries to get ranchers and environmentalists to agree', *New York Times*, 19 February: 8

Freedgood, J., 1994, 'Circuit breakers spread tax relief statewide', *Farmland Update*, Winter(1): 4

Gallagher, M., 1994, 'Florida adopts Everglades pollution law', *Planning*, 60(6): 28–29

Goldfarb, W., 1994, 'Watershed management: slogan or solution?' *Boston College Environmental Affairs Law Review*, 21: 483–509

Lansky, M., 1992, *Beyond the Beauty Strip: saving what's left of our forest*, Tilbury House Publishers, Gardiner, Me

Lapping, M., 1994, '1992–93, the year in review in US rural planning: the promise of change', *Progress in Rural Policy and Planning*, 4: 5–15

National Agricultural Lands Study, 1981, *Final Report*, US Government Printing Office, Washington, DC

Vesterby, M., Heimlich, R. and Krupa, K., 1993, *Urbanization of Land in the United States*, US Department of Agriculture, Economic Research Service, Washington, DC

Williams, T., 1994, 'Incite: whose woods are these?' *Audubon*, 96: 3

# 2 Sustainability, regional planning and the future of New York's Adirondack Park

*Robert J. Mason*

In 1990 the Commission on the Adirondacks in the Twenty-First Century (1990) proclaimed that its proposals for the next century 'combined the greatest wilderness system in the East with working forests and farms, in a way that would continue to provide needed employment for the [Adirondack] Park's 130 000 permanent residents'. By blending the rhetorics of economic and ecological sustainability, the Commission's recommendations laid a hopeful foundation for the future of the Adirondack Park. But in the three years following the release of the Commission's report, no new comprehensive land-use legislation was signed into law. Instead, the report has caused deep political division and prompted renewed state-level efforts to accommodate local interests and concerns. This chapter reviews the course of events leading up to the current stalemate, and then examines the underlying bases of the major conflicts in the Adirondacks. Because the Adirondack Park is part of a worldwide system of 'biosphere reserves' and is regarded within the United States as a model for public management of private lands ('greenline' planning), developments there have important implications for land-use regulatory programs elsewhere.

## Historic context

New York State's history of intervention in the local and regional affairs of its northern mountains is a long one (Graham, 1978: Liroff and Davis, 1981; Terrie, 1985; Heiman, 1988). In 1885 it created the Adirondack and Catskill Forest Preserves, whose lands were to be kept forever in state ownership and 'forever' in a 'wild' condition. In 1892, the 2.8-million-acre Adirondack Park was created, and subsequent boundary expansions have enlarged it to nearly 6 million acres. This vase domain contains private as well as public lands.

In 1895 the 'forever wild' concept was enshrined as a constitutional

*Progress in Rural Policy and Planning*, Volume Five. Edited by Andrew W. Gilg
© 1995 Editors and contributors. Published 1995 by John Wiley & Sons Ltd.

amendment: state lands could not be leased, sold, or exchanged; and their timber could not be sold, removed, or destroyed. The principal impetus for these actions came from progressive political impulses calling for the long-term protection of timber and watersheds to provide for future downstate needs. Open space and recreational amenities were secondary considerations.

Three-quarters of a century later — in 1967 — Laurance Rockefeller put forth his contentious and ill-fated proposal for an Adirondack National Park. In the wake of the controversy, brother Nelson — then Governor of New York State — appointed a Temporary Study Commission on the Future of the Adirondacks. Following on the Commission's recommendations, the Adirondack Park Agency (APA) was created in 1971.

In 1973 one of the most stringent land-use plans in the nation — the APA's Private Land Use and Development Plan — took effect, after the need for such a plan had been demonstrated by proposals for two second-home developments, each in the order of several thousand acres. The 1973 plan created six zones for private Adirondack lands. Within them, development is permitted at various intensities, ranging from virtually unrestricted densities in 'hamlets' (which comprise about 2 per cent of the park's area) to a zoning which would allow one principal building per 42.7 acres in 'resource management' areas, which comprise 53 per cent of the park. 'Industrial', 'moderate intensity', 'low intensity' and 'rural use' are the remaining four classifications.

## The Twenty-First Century Commission

In 1989, in response to widespread concerns within the state's environmental community that the Adirondack Park was threatened with a development crisis, New York Governor Mario Cuomo appointed the Commission on the Adirondacks in the Twenty-First Century. Its membership — as well as associated staff, advisors, consultants, and interns — was weighted heavily toward, though not entirely dominated by, established preservationist interests.

*Recommendations*

The Twenty-First Century Commission report contains some 245 recommendations. While calling for a balance between human and ecological needs, the report states that this balance must be struck in favor of environmental and biological quality. Among the Commission's principal recommendations are the following:

•    The gradual addition of 655 000 acres of lands to the state Forest Preserve, raising state ownership of the Adirondack Park from 42 per cent to 52 per cent as shown in Figure 2.1. For the near term at least, the

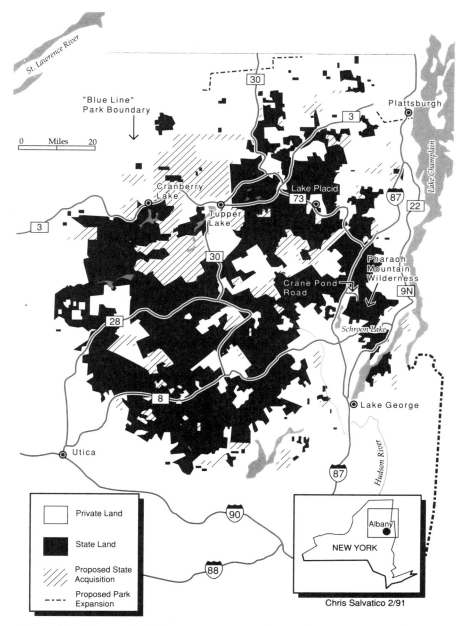

**Figure 2.1**   *Proposals of the Commission on the Adirondacks in the Twenty-First Century*

state would, whenever practical, purchase conservation easements rather than acquire land in full land ownership.

- Two thousand-acre zoning. In resource management areas (now zoned for one principal building per 42.7 acres) and rural use areas (now zoned at one principal building per 8.5 acres) owners would receive one 'structural development right' for up to 2000 acres, and one for each 2000 acres thereafter. A transferable development rights program would be introduced. The objective is containment of the vast bulk of new development in the park's hamlets.
- Imposition of a new and more stringent set of siting and performance standards and negative use incentives to induce local governments to adopt approved land-use plans that incorporate the new standards. As of mid-1994, only 12 of the 105 communities in the Adirondacks had APA-approved plans.
- Imposition of a one-year moratorium on subdivision and development in resource management and rural use zones, as well as within 660 feet of shorelines.
- Creation of a narrow buffer zone around the Adirondack Park's 'blue line' boundary shown in Figure 2.1. Within this new transition zone such activities as agriculture, forestry, wood manufacturing, and tourism would be encouraged.

The report also calls for new investment and economic incentives, among them:

- Incentives and subsidies for economic development, especially to promote forestry and tourism.
- Affordable housing programs. A 1 per cent real estate construction/ transfer tax would provide funds to a newly created Community Development Corporation.
- State-sponsored health care and education programs.
- Expansion of state aid to local governments provided under current formulas.

*Response*

Local response to the report was swift and overwhelmingly negative. New interest groups coalesced, existing groups were revitalized, counter-reports and statements were released, motorcades slowed traffic on the Northway (the Expressway linking Albany and Montreal), and various veiled threats were made. State and national environmental organizations, in contrast, called for quick introduction and passage of legislation to enact the Commission's key recommendations.

But quick action was not forthcoming. Governor Cuomo distanced himself from the report's contentious recommendations, arguing that time was needed to solicit more views on the issues. In late 1990, New York State's voters, concerned about spending and 'politics as usual', broke with their tradition of strong support for environmental initiatives. They rejected the Twenty-First Century Environmental Quality Bond Act, which would have provided $800 million for acquisition of environmentally sensitive lands statewide, much of which would have probably been spent in the Adirondacks.

In 1993 state legislation was enacted that created a greatly scaled-down trust fund, allowing for the purchase of six critical properties in the Adirondacks as well as additional purchases if approved by local governments. Attempts to pass comprehensive new land-use legislation have been stymied by Senator Ronald Stafford, a power broker in the state Senate whose district includes part of the Adirondacks. In 1993, Governor Cuomo introduced legislation that, among other things, would have expanded development opportunities in parts of the park, while further restricting them in shoreline areas and requiring clustering of permitted development in backcountry areas. Although the state Assembly passed a bill that incorporated the Governor's recommendations, the Senate has yet to act. In all likelihood, legislation will eventually be approved, but it will probably fall far short of what the Twenty-First Century Commission recommended.

## The Adirondacks of biosphere reserve and greenline park

Biosphere reserves are designated by the United Nations Man and the Biosphere Program, but their management is the sole domain of national, regional, and local governments. The ideal biosphere reserve has a *core area*, where ecosystem protection is the principal management consideration; a *buffer zone*, which protects the core that it surrounds or adjoins and where a wider range of land uses is allowed; and, finally, a *transition zone*, where human uses ('traditional' land uses in particular) and natural systems are meant to coexist harmoniously (Batisse, 1982, 1985; US Man and the Biosphere Program, 1989; Lucas, 1992; Wells and Brandon, 1992).

The Adirondacks are a distinct physiographic as well as administrative subunit within the 9.9-million-acre Champlain–Adirondack Biosphere Reserve, which includes land in the adjacent state of Vermont. Like many a biosphere reserve, the Adirondack Park does not fit the simple core–buffer–transition model neatly. The core, such as it is, is rather dispersed. More intensive land uses — constituting a buffer zone of sorts — are permitted on the eastern and northern margins of the park, though there are incursions of intensive use into the center, such as at the resort village of Lake Placid. The Twenty-First Century report proposed that a narrow transition zone be added around part of the park's perimeter as shown in Figure 2.1. Within it, agriculture, forestry, wood products manufacturing, and tourism would be encouraged. Although Adirondacks planning and policy making are not explicitly guided by biosphere reserve principles, the issues currently being debated are some of the more contentious and important ones facing biosphere reserves in the industrialized world.

New York State delineated the Adirondack Park in the late 1800s with a 'blue line' on the map. This may be the inspiration for the recent term 'greenline' (Corbett, 1983; Hirner and Mertes, 1986). 'Greenline' parks are relatively large areas of land in a combination of public and private ownership with a mix of such land uses as agriculture, forestry, tourism, and human settlements (Corbett, 1983; Hirner and Mertes, 1986; Belcher and Wellman, 1991; Little, 1992; Mason, 1994). The Adirondack Park has been characterized as an outstanding application of this concept (Little, 1983) and one, that in fact, predates the popularization of the term 'greenline'.

'Greenline' planning tries to do several things: promote viable local economies; protect traditional cultures; ensure ecological preservation; and provide recreation opportunities accessible to major urban populations. These goals, of course, frequently conflict. In the Adirondacks, as elsewhere, it is not clear just when and where one should take precedence over another. Nor does the Adirondack Park conform to the broad intergovernmental management model favored by most greenline advocates: there is virtually no federal involvement and many local governments are, at best, reluctant participants. But the Adirondack program embraces greenline ideals more fully than most other regional land-use programs, with the possible exception of New Jersey's Pinelands National Reserve (Lilieholm and Romm, 1992; Mason, 1992a). Thus, it is an important laboratory as well as model.

*Planning and management issues*

Rhetorically, at least, successful management of 'greenline' parks and biosphere reserves combines ecological protection with the economic wellbeing of the region's residents. In practice, perceptions of these concepts' meanings often differ markedly, even when there is agreement at some level about the need to work towards both objectives. This is the case in the Adirondacks, where some of the deepest and most irreconcilable differences are between the park's permanent residents and those consumers of the landscape and ecology who reside principally outside its boundaries. Several major areas of contention are now described in turn.

*Differing conceptions of nature*   Local politicians in the Adirondacks commonly think in terms of open space rather than wilderness or biodiversity. Nature protection is important to them, but it is important as well that nature be accessible and that it provides direct benefits to humans, for example in the form of recreation, habitat for desirable species of wildlife, harvestable timber, and scenic views. Ecological concepts such as edges, patches, corridors, and succession, theoretically contentious as they may sometimes be among ecologists themselves, are not even part of the lexicon of most local planners and officials. Biodiversity and ecological sustainability, if recognized at all, are not as highly regarded as are economic sustainability, rural ambience, and recreation. 'Wise use' philosophies of resource management also find favor with local politicians, and they would argue that its preservation is accomplished in the two-fifths of the park in state ownership.

Conflicting views are especially in evidence in those areas of the park designated as wilderness. The State of New York stringently regulates such areas, prohibiting most structures, vehicles, motor boats, and airplanes. Two contrasting cases in 1990 illustrated the intensity of feeling about these areas. First, in the Pharoah mountains region, the planned removal of a fire tower was accelerated after its guide wires were cut — an action for which the radical environmental group, Earth First!, claimed responsibility. Second, New York State attempted, unsuccessfully in the face of local protests, to close a road leading into the Pharaoh Mountains Wilderness area. This is an area that local townspeople have used for decades. Although there is no immediate link between the pre-empted closure and release of the Twenty-First Century report, the concurrence of these two actions in 1990 made the Crane Pond area in the southeastern Adirondacks, shown in Figure 2.1, a flashpoint for controversy and a symbol for many Adirondackers of their struggle against a repressive, intrusive state bureaucracy.

*Differing conceptions of crisis*   In his charge to the Twenty-First Century Commission, Governor Mario Cuomo observed that 'Recent developments suggest that we may be entering a new period in the history of the Adirondacks, an era of unbridled land speculation and unwarranted development that may threaten the unique open space and wilderness character of the region' (Commission, 1990, p. iii). In its April 1990 report (Commission, 1990; Miller 1990) the Commission affirmed this cause for concern, pointing out that annual applications for subdivision to the APA more than tripled between 1984 and 1989, and that early projections for 1990 indicated a 72 per cent increase over 1989, already a year with the highest number of applications in the Agency's history. The Commission's supporters also pointed with alarm to the sale by Diamond International Corporation of nearly 100 000 acres of Adirondack forest land. Large land transactions had occurred before, but had been transferred from one timber company to another. In this case, however, Diamond sold the land to a company with speculative interests. Through purchase and acquisition of easements, New York State was able, at great expense, to gain control over 56 000 of the acres sold by Diamond.

The Diamond episode notwithstanding, most juries are still out on the question of whether or not there is a development or subdivision crisis in the Adirondacks, and if so how far back it goes. For example, an analysis by Zinser (1980) concluded that the Adirondacks did not experience a development crisis during the 1970s. A more recent APA study (Adirondack Park Agency, 1990) revealed that as of 1987, over 50 per cent of structures were situated in 'hamlet' or 'moderate intensity' zones, with the remainder lying principally in areas proximate to build-up places or major transportation arteries. However, between 1987 and 1992, according to APA data, three-quarters of new one- and two-family dwellings were built outside hamlets, but many, apparently, went up just outside hamlet boundaries. Also a report commissioned by the Northern Forest Lands Commission, a multistate organization created to study and deal with forest ownership issues in northern New York and New England (Harper *et al.*, 1990), reveals that only 4 per cent of the large tracts of Adirondack forest land sold between 1980 and 1991 were scheduled for development, and that most of the rest were transferred between timber companies or purchased by the state or conservation organizations. Although comprehensive parkwide data on land subdivision are not available, it is apparent that the recessionary trends of the early 1990s have also damped down any tendencies toward rampant overdevelopment, and that there is a lack of compelling evidence to imply a long-term crisis.

Nonetheless, the Twenty-First Century Commission produced its recommendations largely in response to a perceived parkwide 'development crisis', and it will be recalled that while it recommended a balance between human and ecological needs, it also stated that this balance must be struck in favor or environmental and biological quality. Not surprisingly, this general orientation had added fuel to the fires ignited when the Commission was set up in 1989. While environmentalists viewed the economic slowdown of the early 1990s as an opportunity to strengthen regional planning before new crises threatened the Adirondacks, local residents' concerns about their economic plight were only heightened by the prospect of more stringent land-use regulation. Sitting on the fence are local officials who might agree that, though there is a need for further regulation, they are also inclined to temper or withhold their support in a recessionary climate. To them, the 'crisis' is one of too little, not too much, economic activity.

*Differing conceptions of appropriate economic development*    One vision of the ideal 'greenline' park or biosphere reserve has its residents employed in traditional, renewable resource-based activities such as farming, forestry, fishing, and wood product manufacturing as well as such seemingly benign pursuits as craftwork and tourism. These are all important activities for the Adirondacks and there are strong arguments to be made for public investments and incentives to encourage them. At the same time, many local officials and residents do not want to preclude industrial development (limited though the potential may be) or less desirable public investments. Indeed, not only manufacturing facilities but also state prisons and federal military installations are sought after by many Adirondacks towns.

Although the extent to which the Adirondacks population constitutes a traditional society is highly arguable, West and Brechin's (1991, p. 380) question is still poignant: 'What happens when a traditional society wishes to adopt more modern means of harvesting resources?' Almost all would agree that the Adirondack economy must adapt to an era in which the region's role as a natural resource hinterland continues to diminish in importance. The debate is over how to achieve this transition and what mix of economic activities to allow and encourage. At the biocentric extreme, the national 'Wildlands Project' has put forth a proposal calling for an eventual ban on motor vehicles throughout the Adirondacks. Some residents would be relocated outside the park, those who remain would be employed in wilderness restoration projects; forest crafts and other ecologically sustainable pursuits (Medeiros, 1992).

The more moderate among the locally based parties to the debate recognize that there is a demand to be met for environmentally correct tourism, craftwork, and the like, but argue that the economic transition will not come easily. The logger or building contractor, for example, may feel diminished by having to attend to visitors' needs. As one resident stated in an open forum: 'There are not enough duvets to fluff to employ all of us.' In addition, tourism's benefits, often hailed as the economic savior for regions like the Adirondacks, are distributed unevenly. There is little to attract tourism to some parts of the region and an abundance of amenities elsewhere. The region is not uniform: physically; socially; or economically.

*Equity considerations*  The report of the Twenty-First Century Commission only skirts the basic questions of equity that have pervaded Adirondack life for the past century. Before the Second World War there were essentially two classes of Adirondackers: the region's full-time residents, most of whom struggled to make ends meet; and the select stratum of wealthy landowners and other part-time residents. After the war, the region became much more accessible to middle-class cottagers and other visitors, but the comparative lot of native Adirondackers did not markedly improve.

On the one hand, the report of the Twenty-First Century Commission calls for regional health care, education, affordable housing, and other state aid programs. But on the other, it perpetuates — indeed may widen — the gap between rich and poor. By calling for 2000-acre zoning outside the hamlets it keeps the open spaces of the Adirondacks accessible to the privileged elite but off-limits to those of lesser means. The lower-income, local population will have to expand principally within the park's hamlets.

At a broader scale, many local residents feel that they are being unfairly singled out by the state of New York. Why, they ask, are stringent land-use regulations applied only in the Adirondacks? While there are, of course, compelling arguments for singling out places such as the Adirondacks for special protection, even those local residents supportive of regional planning feel that their interests have not been properly represented.

*What is local participation and how important is it?*   Local participation and sustainability seem to be inextricably linked in the literature on biosphere reserves and other protected areas where humans are part of the landscape. In practice, however, true local control and bottom-up planning are often rejected in favor of token measures (Wells and Brandon, 1992). This is the case in the Adirondacks. Advocates of preservation generally support the idea of local participation in park decisions but oppose increased sharing of power in practice. If they could be assured that environmentally correct decisions, or at least decisions no more detrimental than those currently made, would be forthcoming, then they would probably support greater sharing of power. As it is, such steps are regarded as risky, far too risky to be openly embraced.

There are no locally appointed members of the Adirondack Park Agency, and although local officials and private organizations have argued for local representation they have been and continue to be strongly opposed by environmental organizations. What has instead taken place is a much more limited form of power sharing. Some recent actions have allowed local residents and their representatives to get a legitimate foot in the APA's door. John Collins, a fifth-generation Adirondacker, was recently appointed Executive Director, and in 1993, Governor Cuomo appointed three full-time local residents to the Agency. An open issues forum at monthly APA meetings also allows airing of local concerns and an APA task force that recently reported on ways to expedite APA operations and simplify procedures included several local representatives. At the same time, however, recent state budget cuts have greatly restricted the financial and technical assistance that the APA can offer to local governments. Moreover, the Governor's legislative proposals have not included any requirements, as demanded by many Adirondack interest groups, similar to those in place in New Jersey's Pinelands, for locally appointed representatives to the regional agency.

Two principal arguments could be made in favor of such power-sharing. One is a political equity argument, namely that local residents of the Adirondacks should have the same political power and rights to self-determination as other residents of New York State. This argument is made repeatedly by politicians from the region. A second, perhaps more compelling, argument is of a more practical nature. It has been pointed out in the planning literature (Hahn and Dyballa, 1981) that strong state-level support for critical-area protection can overpower local opposition, especially in a populous state where a tiny fraction of the population resides in the protected area. That argument applies to the original 1973 Private Land Use and Development Plan for the Adirondacks. But legislation introduced since the 1990 release of the Twenty-First Century report has failed to progress through the state legislature and Governor Cuomo has distanced himself from many of the recommendations of the Commission he appointed. This is an indication of how effective local opposition can be when politicians from the affected area built much of their political career around land-use issues (Mason, 1992b). This is the case in New York, where Ronald Stafford, who is Deputy Majority Leader of the state Senate and whose district includes part of the Adirondacks, has blocked legislation that would more stringently regulate the Adirondacks.

Then there is the question of what constitutes local participation. For example, some literature reviews interpret local quite generously (West and Brechin, 1991). Indeed, this is the case for various reviews of the New Jersey Pinelands experience (see, for example, Little, 1992), where some observers regard all New Jersey interest groups, as well as those from nearby parts of New York and Pennsylvania, as local Pinelands interest groups. Although the Adirondacks are more physically and politically isolated, there are still those who would argue that the Adirondack Council is a local participant in the planning process, although it is composed largely of downstate New York City area interests, and while most of its meetings are held in New York City.

*Can economic incentives bring about local compliance with regional plans?* Accommodation of local interests is essential to the success of regional land-use programs. The most prominent of the 1970s 'quiet revolution' programs had made administrative and economic concessions by (if not well before) the mid-1980s (see Popper, 1981; DeGrove, 1984). In the 1970s the APA simplified its review procedures for projects, substituted civil for criminal penalties for violators of the regulations, and provided greater planning assistance to localities. The 1980s witnessed greater APA emphasis on economic development.

Many Adirondackers, including some rather vocal opponents of the APA's creation, had by the late 1980s come to grudgingly accept the added layer of regulation. They realized that it was not going to go away, and that concessions had been made to local interests. These concessions helped buy local complacency, thus obviating to some degree the need to develop trust and cooperative working relationships with local residents. Such has indeed been the case in New Jersey's Pinelands National Reserve (Mason, 1992a).

But, as recent Adirondack experience demonstrates, there are limits to this strategy. While many of the incentives and inducements proposed by the Twenty-First Century Commission are welcomed locally, they are entirely insufficient as substitutes for the stringent new regulations that are proposed by some. Furthermore, although some of the Commission's work is welcomed it has failed, by and large, to enthuse local residents to be on the side of the Commission. In addition, the Commission's 1990 report devoted little attention, overall, to basic economic issues, and recent efforts to bring about greater local involvement in Adirondack decision making came only after years of stalemate that followed the report's release.

## Conclusion

During its first decade the Adirondack Park Agency pursued principally a preservationist mission. Most of its second decade, in contrast, witnessed greater conciliation with local interests and a higher priority for economic development. But the 1990 report of the Commission on the Adirondacks in the Twenty-First Century, conceived, developed, and released with little local participation (even the APA was excluded from the deliberations) and coming as it did during a recession, provided fresh fuel to opponents of regional planning. It brought renewed animosity towards the Adirondack Park Agency, effectively ending what grudging acceptance and trust the Agency had built over the previous ten years.

This entire experience points towards the inherently contradictory nature of protected area management principles, at least as actually practised. Managers strive for the implementation of a broad vision, making the process inherently top-down. Yet at the same time, local participation and economic sustainability are promoted. Based on his study of the Buffalo National River in Arkansas, Sax (1984) raises compelling questions about the respective roles of the national interest and local autonomy and diversity. The relationship between local and larger-level interests will perhaps never be an easy one. Much as we may wish them to be, economic and ecological sustainability will not always be compatible; thus, as Sax acknowledges, there are times when the larger interest must override local concerns. The challenge for places like the Adirondacks is in deciding just when and where particular ecological axioms *are* to be regarded as inviolable. The Twenty-First Century report itself is quite clear: ecology must come first.

But there are few hard and fast rules, even for the management of a purely ecological preserve. We are engaging in an experiment, seeking, perhaps, a dynamic steady state that has socio-cultural as well as ecological implications. A wider, shared recognition of the experimental nature of this enterprise, without rhetorical invocations of crisis, may help bring about more meaningful dialogue. Through a painful process of conflict and political compromise, it appears that there has been some very recent, and limited, movement in this direction in the Adirondacks. Indeed, there is at least some small common ground: local politicians, by and large, are as fearful as are their statewide counterparts that fringe individuals, on both the property rights right and the ecocentric left, will continue to stir up local hysteria. That the Adirondacks have been ravaged and have recovered some semblance of ecological wholeness and security within the space of a century should give hope to all involved that a workable regional vision can be found for the succeeding century.

# References

Adirondack Park Agency, 1990, *Development Patterns in the Adirondack Park: 1967–1987*, Adirondack Park Agency, Ray Brook, New York
Batisse, M., 1982, 'The biosphere reserve: a tool for environmental conservation and management', *Environmental Conservation*, 9: 101–11
Batisse, M., 1985, 'Action plan for biosphere reserves', *Environmental Conservation*, 12: 17–27
Belcher, E.H. and Wellman, J.D., 1991, 'Confronting the challenge of greenline parks: limits of the traditional administrative approach', *Environmental Management*, 15: 321–8
Commission on the Adirondacks in the Twenty-First century, 1990, *The Adirondack Park in the Twenty-First Century*, State of New York, Albany, New York
Corbett, M.R. (ed.), 1983, *Greenline Parks: land conservation trends for the 1980s and beyond*, National Parks and Conservation Association, Washington, DC
DeGrove, J.M., 1984, *Land Growth and Politics*, American Planning Association, Washington and Chicago
Graham, F.J., Jr, 1978, *The Adirondack Park: a political history*, Alfred A. Knopf, New York
Hahn, A.J. and Dyballa, C., 1981, 'State environmental planning and local influence: a comparison of three natural resource management agencies', *American Planning Association Journal*, 47: 324–35
Harper, S.C., Falk, L.L. and Rankin, E.W., 1990, *The Northern Forest Lands Study of New England and New York*, US Department of Agriculture, Forest Service, Rutland, Vermont
Heiman, M.K., 1988, *The Quiet Evolution: power, planning, and profits in New York State*, Praeger, New York
Hirner, D.K. and Mertes, J.D., 1986, 'Greenlining for landscape preservation', *Parks and Recreation*, 21(11): 30–34 and 59
Lilieholm, R.J. and Romm, J., 1992, 'Pinelands National Reserve: an intergovernmental approach to nature protection', *Environmental Management*, 16: 335–43
Liroff, R.A. and Davis, G.G., 1981, *Protecting Open Space: land use control in the Adirondack Park*, Ballinger, Cambridge, Mass.
Little, C.E., 1983, 'The national perspective: greenline parks', in *Proceedings: Greenline and Urbanline Barks Conference*, New York State Department of Environmental Conservation, Albany, New York: 3–5
Little, C.E., 1992, *Hope for the Land*, Rutgers University Press, New Brunswick, New Jersey
Lucas, P.H.C., 1992, *Protected Landscapes: a guide for policy-makers and planners*, Chapman & Hall, London
Mason, R.J., 1992a, *Contested Lands: conflict and compromise in New Jersey's Pine Barrens*, Temple University Press, Philadelphia
Mason, R.J., 1992b, 'Defining and protecting rural environments in the US', in Bowler, I.R., Bryant, C.R. and Nellis, N.D. (eds), *Contemporary Rural Systems in Transition*. Vol. 2: *Economy and society*, CAB International, Wallingford, 129–40

Mason, R.J., 1994, 'The greenlining of America: managing private lands for public purposes', *Land Use Policy*, 11: 208–21

Medeiros, P., 1992, 'A proposal for an Adirondack primeval', *Wild Earth*, special issue: 32–42

Miller, P.N., 1990, *Subdivision and Development Trends: extent and location (the Adirondack Park in the twenty-first century technical report 38)*, State of New York, Albany, New York

Popper, F.J., 1981, *The Politics of Land Use Reform*, University of Wisconsin Press, Madison, Wisconsin

Sax, J.L., 1984, 'Do communities have rights? The national parks as a laboratory of new ideas', *University of Pittsburgh Law Review*, 45: 499–511

Terrie, P., 1985, *Forever Wild: environmental aesthetics and the Adirondack Forest Preserve*, Temple University Press, Philadelphia

US Man and the Biosphere Program, 1989, 'Biosphere reserves: what, where and why?' *Focus*, 39(1): 17–19

Wells, M. and Brandon, K., 1992, *People and Parks: linking protected area management with local communities*, The World Bank, The World Wildlife Fund, and US Agency for International Development, Washington, DC

West, P.C. and Brechin S.R. (eds), 1991, *Resident Peoples and national parks: social dilemmas and strategies in international conservation*, University of Arizona Press, Tucson, Arizona

Zinser, C.I., 1980, *The Economic Impact of the Adirondack Park Private Land Use and Development Plan*, State University of New York Press, Albany, New York

# Section II:
# United Kingdom

edited by
*Andrew W. Gilg*

# Introduction
## *Andrew W. Gilg*

This year's UK section contains two chapters: first, the 'annual review' of rural planning in the UK; and second, a chapter on sustainability, equality, and pluriactivity by a team of researchers who have been looking at how these concepts are developing in Northern Ireland in the context of a European-wide research effort. They conclude that the greening of attitudes to farming by both policy makers and farming households could lead the way towards sustaining not only a modified form of pluriactive family farming but also a more diverse wildlife and landscape. At the moment these effects are perhaps unintentional, and so the authors argue that the rural policy milieu should begin to offer more distinct policy signals combining socio-cultural change with environmental conservation and land management.

The section is, however, dominated by the first chapter, the 'annual review'. This year the annual review has been substantially restructured into an essay style divided by rural planning activity rather than by legislative category. The following structure has been employed:

- Changes affecting all rural planning activities
- Agriculture as a productive industry
- Agriculture and the environment
- Forestry
- Water and the coast
- Conservation and recreation
- Town and county planning
- Social and economic issues

The year saw a number of themes, most notably the gradual greening of rural policy via the series of four White Papers published in January 1994, which set out the government's programme for achieving the commitments it had made at the 1992 Rio Conference on the Environment and Development. However, spending on agriculture as a productive industry continued to grow, and at some £3000 million it still dwarfs expenditure on agriculture and the environment, which accounts for only some £100 million. Nonetheless, too many academics still preface their writings with remarks about a

*Progress in Rural Policy and Planning*, Volume Five. Edited by Andrew W. Gilg
© 1995 Editors and contributors. Published 1995 by John Wiley & Sons Ltd.

'post-productivist agriculture'. One day on the farm would make them realise just how silly such a phrase is.

Forestry was not only the subject of one of the Rio White Papers, but also the subject of a major review of forestry policy, which reprieved the Forestry Commission from privatisation. The Forestry Commission for all its faults, continues to be a potential model for other rural planning agencies. However, its lack of democratic accountability remains a thorn in its flesh. This, however, is a problem with most rural planning agencies who are run by non-elected nominees from whichever political party is in power. It is time that patronage was replaced by democracy.

The Countryside Commission and English Nature did not merge after all, and elsewhere County Councils were reprieved in great numbers. The Conservative Party, once so committed to radical reform, seemed to lose its nerve during the year, and the retention of the Post Office in the public sector at the end of the year was one of many examples of the status quo being maintained. Nonetheless, the year ended with Bills being published which would honour some long-standing commitments, notably the 1992 election pledge to enact the recommendations of the 1991 report on the future of National Parks, notably the creation of unitary National Park Authorities.

The most significant announcement, however, may be the joint decision by DOE and MAFF to publish a White Paper on the Countryside in 1995. At long last we may have some guidance about how different activities in the countryside may be reconciled, and a framework within which each rural planning activity can operate and relate to other organisations.

# 3 Annual review of rural planning in the United Kingdom: autumn 1993–autumn 1994

*Andrew W. Gilg*

## Introduction

This is the 14th annual review I have written for PIRPAP and its forebears, but it is a radically different style of review in that it takes the form of an essay rather than a structured review focused around the type of publication. The change has been necessitated by the need to keep PIRPAP to 80 000 words, and has been made possible by modern word processing power which allows the former style of review to be edited and pasted to the desired length. If readers wish me to return to the former style, please let me know. In spite of the change, I am still collecting information about rural planning in the style of a voracious vacuum cleaner, and this year's review has been boiled down from a 3-foot pile of raw material and several disks. If any reader would like material from this source, the literature list, for example, is available in a similar style to that of previous years from the editor.

## Changes affecting all rural planning activities

*Sustainable development*

The main event of the year was the government's response to the 1992 Rio conference on the world environment and its flagship concept of sustainable development. The response was contained in four White Papers published in January 1994: Cms 2426 to 2429. Three of these are considered here, the fourth — 2429 — on forestry is considered in the forestry section on p. 55. The first White Paper — Cm 2426, Sustainable Development: the UK Strategy — builds directly on the three White Papers on the environment published since 1990 under the generic title of *This Common Inheritance*. In common with these White Papers the 'UK Strategy' is well presented, lucid

*Progress in Rural Policy and Planning*, Volume Five. Edited by Andrew W. Gilg
© 1995 Editors and contributors. Published 1995 by John Wiley & Sons Ltd.

and comprehensive. However, also in common with *This Common Inheritance* the proposals made in the *UK Strategy* are bland, and in general build on existing policies rather than introducing radically new ones. This is highlighted by the introductory boxes to each of the policy chapters which use the common headings of A Sustainable Framework; Trends; Problems and Opportunities; Current Responses; and The Way Forward. In spite of these reservations, the White Paper marks another step towards a genuinely green environmental policy for the UK, since in the document the government has set out the framework for further progress, which is to be achieved via the government's flagship mechanism, the market. In the long term, though, as the White Paper acknowledges, government can only do so much without the general acceptance of the public, and so the aim must be to 'increase people's awareness of the part their personal choices play in delivering sustainable development and thus to enlist their support and commitment over the coming years' (p. 19). As an aid to this process the government set up five initiatives to study possible ways forward.

However, there were few indications how these changes would actually be achieved, except for a general presumption against car-borne commuting and a long-overdue backlash against the roads lobby. Not surprisingly, the general reaction was muted in that the general sentiments of the White Paper were welcomed, while the lack of detailed commitments was criticised. Typical comments were: 'more style than substance'; 'one cheer . . . for a grandiose discussion paper that contains few targets'; and even 'Retreat from Rio' which headlined a most useful series of articles in *Ecos*, 14, pp. 61–75. Meanwhile, the House of Lords Select Committee on Sustainable Development heard evidence from several interested parties including the DOE, academic experts like David Pearce, and the energy industry in HL 66 i–viii (93–94).

Not surprisingly, most of the main groups had already published their own views in advance of the White Paper in 1993. For example, the Countryside Commission had issued *Sustainability and the English Countryside*, Scottish Natural Heritage had published *Sustainable Development and the Natural Heritage*, CPRE had published *Preparing for the Future* and *Sense and Sustainability*, and the World Wide Fund for Nature UK had produced a shadow sustainability plan in *Changing Direction: towards a green Britain*. On a wider scale David Pearce and others extended their well-known views on the use of environmental economics in *Blueprints Three* and *Four: Measuring Sustainable Development* (1993) and *Sustaining the Earth* (1994). Finally, Holmberg, Thompson and Timberlake in *Facing the Future: Beyond the Earth Summit* (1993) and Williams and Haughton in *Perspectives towards Sustainable Environmental Development* (1994) provided an invaluable briefing and a comparison of orthodox and environmental economics. However, just before the White Papers were published the government also heralded the long-awaited completion of the Uruguay round of the GATT trade talks in December 1993 (Cm 2579) on the grounds that it would lead to more trade and production — sentiments that some would say were in direct contradiction to a commitment to genuine sustainable development.

Perhaps to counter such accusations the government also published two documents from the department of the Environment. The first of these, *Making markets work for the environment* (1993), claimed that economic instruments compared with other policy instruments encourage cost-effectiveness; induce innovation; provide flexibility; generate information; and contribute additional public revenue. Seven instruments are identified including emissions charges and tradeable permits, which can be assessed against seven criteria: environmental effectiveness; resource costs; administrative costs; public revenues; innovation; competition and competitiveness; and fairness. The second document was concerned with *Environmental appraisal in government departments* (1994). International guidance was provided by the OECD in *The distributive effects of economic instruments for environmental policy* (1994). The White Paper was also preceded by the findings of a major initiative which brought together a wide cross-section of interested parties under the heading of the Centre of Science and Technology which set out its findings in the £133 report *Environmental Foresight Project*. This concluded that the UK's future environmental agenda will be shaped by world growth trends — themselves dominated by population growth, urbanisation and a breakdown of environmental borders.

The second White Paper, Cm. 2427, *Climate change: the UK programme*, set out a programme of measures designed to meet the UK's commitments under the Climate Convention agreed at Rio. Most of these focused on the need to cut emissions of carbon dioxide in the year 2000 to 1990 levels, mainly by taxing energy use above the rate of inflation, notably by raising petrol taxes by 5 per cent more than the rate of inflation each year. Beyond 2000 the White Paper considers ways of curbing the substantial growth of traffic and other energy-related activities predicted for 2025. These include fiscal measures, planning policies designed to minimise travel, and technical improvements. Another significant contribution could be made by afforestation which provides a carbon sink to soak up carbon dioxide from the atmosphere, and in the future will provide a source of renewable energy that is largely carbon-neutral.

The third White Paper, Cm 2428, *Biodiversity: the UK action plan*, sets out 59 specific commitments to forward action and emphasised what can be done at the local level in conserving local biodiversity. At the national level the government committed itself to conserve and, where possible, enhance: the overall populations and natural ranges of native species; internationally important and threatened species, habitats and ecosystems; species, habitats and natural and managed ecosystems that are characteristic of local areas; and the biodiversity of natural and semi-natural habitas where this has been diminished over recent decades. The White Paper was welcomed as the most conservation committed of the series, albeit one that was still too cautious, especially when compared with other blueprints, for example, from the Joint Nature Conservation Committee, *Action for biodiversity in the UK* (1993) and a consortium of groups including Friends of the Earth, the World Wide Fund for Nature and the RSPB in *Biodiversity Challenge — An Agenda for Conservation Action in the UK* (1993). In addition to the White Paper the government also ratified the Biodiversity Convention in June 1994.

## This Common Inheritance

The fourth White Paper in this series (Cm 2549) was published several months late in May 1994 and without the publicity surrounding the first three. It claimed that encouraging progress had been made, and that sustainable development is now the touchstone of policy. As before, each of the 657 commitments are divided into summary columns showing: a summary of previous White Paper commitments; action to date; and commitments to further action. In conclusion, the collected papers are an excellent source of information and *aide-mémoire*.

## Overall spending plans

Plans for overall government spending announced in the new unified budget in November 1993 were published in several places, as shown in Table 3.1, which updates and amends Table 8.1 in Volume Four. Further details are provided in HC 909(92–93), HC 276(93–94) and Cm 2481.

## Political developments

In a general move to deregulate society the government introduced the Deregulation and Contracting Out Bill 1994. This Bill will allow ministers to change primary legislation by statutory instruments, rather than by primary legislation. However, none of the specific pieces of legislation mentioned in the Bill were directly relevant to rural planning, although the government did single out seven areas where general planning procedures could be amended under the Bill. Nonetheless, the Bill marked a continuing trend to deregulation which in due course may yet have a significant impact on rural planning.

Another trend has been to divide government into a small core of policy makers, surrounded by a host of executive agencies and quangos. Some 60 per cent of public servants now work in agencies. This trend has been greeted with unease by civil servants and the public alike. The process, which began in 1988, is known as 'next steps' and is analysed in Cm 2430, *Next steps:*

Table 3.1   *Public expenditure on rural planning in the 1990s*

| Department/Ministry (£million) | 1992–3 | 1994–5 | 1996–7 |
|---|---|---|---|
| Department of the Environment(Cm 2507) | | | |
| Environmental Protection | 251 | 303 | 285 |
| Countryside and wildlife | 118 | 142 | 148 |
| National Rivers Authority | 74 | 67 | 57 |
| Countryside Commission | 42 | 46 | 43 |
| National Parks and Broads | 16 | 17 | 16 |
| English Nature | 38 | 40 | 40 |
| Rural Development Commission | 16 | 31 | 40 |
| Ministry of Agriculture(Cm 2503)* | | | |
| UK expenditure on market support | 1699 | 2592 | 2866 |
| Receipts from EAGGF | 1525 | 2728 | 2939 |
| Major capital grants | 31 | 28 | 19 |
| Environmentally Sensitive Areas | 11 | 31 | 43 |
| Arable area payments | 0 | 1097 | 1274 |
| New agri-environment schemes | 0 | 1 | 19 |
| Cereals | 262 | 116 | 78 |
| Sugar | 130 | 101 | 101 |
| Beef and veal | 349 | 475 | 567 |
| Milk and milk products | 212 | 235 | 211 |
| Sheepmeat | 339 | 487 | 541 |
| Research and development | 143 | 137 | – |
| Forestry Commission(Cm 2514) | 99 | 95 | 94 |
| Scotland(Cm 2514) | | | |
| Agriculture | 311 | 512 | 572 |
| Highlands and Islands Enterprise | 62 | 58 | 61 |
| Scottish Natural Heritage | 35 | 39 | – |
| Wales(Cm 2515) | | | |
| Agriculture | 167 | 234 | 275 |
| Countryside Council for Wales | 17 | 20 | – |
| Development Board for Rural Wales | 17 | 16 | – |

*Notes*: All figures have been rounded down.
Only expenditure above £10 million shown
* Some of these totals include double counting

*agencies in government: review 1993*. This document examines both the programme as a whole and the progress of the 93 agencies so far devolved. Out of these 93, there are five with rural planning interests: ADAS, Cadw (Welsh Historic Monuments); Historic Scotland; the Intervention Board; and the Scottish Agricultural Science Agency.

*Rural White Paper*

Perhaps the most significant development in the year was, however, the announcement at the Conservative Party Conference in October 1994 that the government would be publishing a White Paper on the countryside, to be jointly prepared by MAFF and the DOE, some time during 1995. The paper would examine the economic, social and environmental changes taking place in our countryside today, building on the principles set out in the Sustainable Development Strategy. It was intended that it would cover all aspects of living and working in and enjoying the countryside. For example, it would examine how to promote economic development while protecting what makes our countryside unique; the dynamic relationship between urban areas and the countryside; the future role of the CAP; and how government policies across all departments impact on rural areas. It is planned to devote a major section of Volume Six to the White Paper.

Finally, both the Labour and Liberal Parties published policy papers on the countryside and the environment offering greener alternatives to government policies. For example, the Labour Party *In Trust for Tomorrow* proposed a halt to road building, increasing forest cover by 50 per cent by the year 2010, and access to open country. The Liberal Democrats outlined their proposals in *Reclaiming the Countryside*. However, the election of the amazingly charismatic and highly electable Tony Blair as leader of the Labour Party in July 1994 could be the most significant event of the year, for he promised to be not only the first politician in 15 years able to turn back the Tory tide but one who offered a really radical but nonetheless centre-view vision of how Britain could be so better governed and its countryside planned and managed. Thirty years too late Britain has found its own Kennedy.

## Agriculture as a productive industry

*International agreements*

The main development during the year was the conclusion of the Uruguay Round of the GATT world trade talks after 7 years. Agreement was finally reached in December 1993, and for the first time agriculture was included in the multilateral trade rules under the GATT which were expected to come into operation in July 1995. In agriculture the agreement was based very closely on the Blair House deal struck in November 1992 which contained four key changes: first, a 20 per cent cut in farm subsidies; second, a 36 per cent cut in import tariffs; third, a 36 per cent cut in export subsidies; and fourth, a 21 per cent cut in the volume of subsidised exports. The revised 1993 deal phased the cuts over a longer period, namely to the year 2000, instead of 1999, and changed some of the base periods on which the cuts were based. The essential move towards a less distorted world market was thus fully maintained.

For British farmers it was forecast that the agreement would mean a wheat price of around £85 per tonne by the end of the century, and set-aside rates of 20–25 per cent if only a modest 1 per cent rise in yields was achieved. The *Farmers Weekly* of 24 December 1993 forecast fewer but larger farms with less time and money to spend on environmental projects.

*The European dimension*

Agreement over farm prices for 1994/5 was reached in July 1994, which although it did not alter the essence of the May 1992 reforms to the CAP, nonetheless increased the projected farm budget by around 5 per cent or 293 mECU extra in 1995. However, this rise was not enough to offset the projected cuts for 1995 agreed in 1992, and the UK government argued that 'substantial savings' of around 1700 million ECU would still be made in 1995. A discussion on the original price proposals is contained in HC 48–xii(93–94). The UK government also managed to prevent a 1 per cent cut in the milk quota for 1994/5, and gained agreement that there would be no quota cut in 1995/6.

The UK green rate changed three times in July and August 1994 with devaluations of 1.23 per cent, 1.49 per cent and 0.74 per cent, leading to agricultural support price rises of 1.25 per cent, 1.51 per cent and 0.74 per cent. The final rate was £0.953575 = 1 ECU which once again helped farmers offset the price support reductions contained in the May 1992 reforms. The European Commission also introduced a replacement for the 1991–92 apple orchard grubbing scheme, and extended it to part orchards. The rate of grant was set at £4600 per hectare.

The House of Lords Select Committee on the European Communities (HL 57(93–94)) warned that a radical reform of the CAP would be a prerequisite to the former Communist countries in Eastern Europe joining the EU in the future. They also reiterated their 1991 calls (HL 79(90–91)) for a reduction in CAP prices to world market prices; unfastening agricultural policy mechanisms from those relating to regional, social, economic or environmental policies; and ensuring that compensation payments to farmers for adjusting to lower price support should be transitional, and be decoupled from production decisions.

More immediate concerns were addressed by the House of Commons Select Committee on European Legislation in HC 878(92–93) and HC 48(93–94), notably on the operation of milk quotas. Returning to the longer term, September 1994 saw the Ministry of Agriculture set up a new team to prepare for the next stage of CAP reform in 1996.

Finally, a summary of changes in European agricultural policy between January and December 1993 are provided by Cm 2369 and 2525, in particular: details of the 1993/4 price fixing; changes to the agrimonetary system in line with the modified ERM; and new rules for set-aside.

*The Ministry of Agriculture and its agencies*

The fourth Departmental report for the Ministry and the Intervention Board was published in February 1994 (Cm 2503). This excellent publication highlighted several major developments, including new environmental measures; enactment of the Agriculture Act 1993; market support developments; the further implementation of CAP reform; and the new agrimonetary system. Expenditure was forecast to rise from £3861 million in 1994–5 to £4170 million in 1996–7. Within these totals, £2642 million and £2917 million were to be spent on the CAP, and £1078 million and £1117 million on domestic policies. It can thus be seen that CAP expenditure is planned to rise faster than domestic expenditure. Total spending on environmental schemes was set to rise from £81 million in 1994–6 to £93 million in 1996–7, with a rise in expenditure of £18 million in the new agri-environment schemes, to some extent being offset by falls in expenditure on Capital Grants, the initial Five Year Set-Aside Scheme, the Hill Livestock Compensatory Allowances Scheme, and in research spending.

A commentary on the Report was produced by the House of Commons Agriculture Committee in HC 434(93–94). This expressed concern that significant changes expected in the format of the Report as a result of Treasury-inspired changes to the structure of the Estimates could lead to some loss of detail. This is because information formerly published in the Estimates will in future be placed in the Report, causing not only pressure on space but by regularising table subheadings will also lead to a loss of comparative data. Finally, the Committee called for the inclusion of performance targets and indicators in future Reports.

The government responded in HC 926(92–93) to eight points made in a previous report by the Agriculture Committee (HC 636(92–93)) on the MAFF annual report for 1993. In particular, it reacted to criticisms of the Ministry's aims established in 1990, and agreed that some of them needed rethinking. However, although the government acknowledged that the aims do not correspond with the structural divisions within the ministry, they argued that it seemed right to maintain the distinction between them, since the aims highlighted the motivations behind the ministry's work. These comments, however, confirmed the suspicion that these Reports are produced for cosmetic reasons, rather than an as an accurate reflection of what the ministry really does, promote production.

Meanwhile the final accounts of the Intervention Board for the six months ended 31 March 1992 (HC 44(93–94)) marked the transition to its successor body, the Intervention Board Executive Agency which took over all aspects of implementing the financial aspects of the CAP in the UK as from 1 April 1992.

ADAS in its Annual Report (HC 580(93–94)) reported that in 1993/4 it had raised the recovery of costs from charged advice to 53 per cent from 45 per cent in 1992/3, and that the target for 1994/5 was 63 per cent. This was against the background of rumours that it was being prepared for privatisation or a management buy-out. Plans for the research strategy on which ADAS relies in the long term were set out by MAFF in their *Research Strategy 1994*. This outlined plans to spend £128 million: £55 million on ways to protect the public; £49 million on improving the economic performance of the industry, £20 million on enhancing the environment; and £4 million on protecting farm animals. Later in the year, in June 1994, the minister announced plans to treble the £4.5 million budget for the LINK: Technologies for Sustainable Farming Systems programme to £12 million. However in December 1993, the Priorities Board for Research and Development in Agriculture was dissolved after making recommendations about how level funding should best be distributed among different areas of agricultural research. Research spending in Scotland was set out in HC 452(93–94).

Also in Scotland the Crofters (Scotland) Act 1993 consolidated certain enactments relating to crofting and gave effect to recommendations made by the Scottish Law Commission. Other changes to crofting were made in SIs 1013 and 1014/94. Other detailed changes to agricultural policy were provided by SIs 2038 and 2039/94 and 249/94.

In addition to spending money on agriculture, MAFF is also a regulator. During 1994 the minister announced various proposals aimed at deregulating agriculture wherever possible. The review process will be divided into five stages. First, it will ask if the regulation is needed, second, it will try to simplify those regulations that remain, third, it will resist new burdens being imposed, especially from the EU, fourth, it will target action on risks, and fifth, it will inform and consult. In July a number of measures were proposed, which included: the use of set-aside land for new woods; relaxing other set-aside rules; and introducing greater flexibility into the transfer of milk quotas.

Finally, in two speeches in February and March, the minister emphasised the continuing importance of agriculture to the economy, the balance of trade, and the environment. The most important priorities and policy aims were then set out as to further reform the CAP by reducing support prices and bringing supply and demand into balance; deregulating the domestic market; removing unnecessary regulation; fighting for a level playing field in Europe; promoting better marketing of produce; encouraging environmentally beneficial farming practice; supporting improvements via research; sustaining agriculture in vulnerable and remote areas; and integrating environmental aims more closely with agricultural policy.

The minister also set out to remind the nation that agriculture is still a keystone of the rural and national economy, in spite of ill-informed academic commentaries that agriculture has reached a post-productionist phase. For example, the food chain employs 14 per cent of the nation's workforce, and accounts for 9 per cent of the country's GDP, and 6.6 per cent of total exports. However, farming still only produces 73 per cent of the country's food we are able to grow, and it should be expanding, not contracting, in a properly planned world.

## Arable farming

Set-aside continued to be the main issue as the various schemes that have been introduced since the late 1980s continued to evolve. For example, in December 1993 the UK government managed to convince the EU that the set-aside requirement for mixtures of rotational and non-rotational set-aside would be 18 per cent, rather than the 20 per cent proposed. The original five-year scheme was also extended to allow farmers to register for another five years, thus continuing any conservation benefits. Detailed rules for both schemes were issued in March 1994, and revised regulations for Arable Area Payments were published in March 1994 in SI 947/94.

In May 1994 MAFF published figures which showed that claims had been made for payment for 2.77 million ha of growing crops, mainly cereals, and for 492 000 ha of set-aside land. In addition, claims were made for 209 000 ha under the simplified scheme.

In September 1994 new payment rates for 1994/5 were published as shown in Table 3.2. This distributed the same 'pot of gold' in a slightly different way as can be seen when Table 3.2 is compared to the figures given for 1993/4 in Volume Four on p. 109. However, it was expected that payments for maize would fall because of a 70 per cent overshoot on the separate maize base area. Finally, in October 1994 the Council of Ministers provisionally agreed to cut the set-aside rate from 15 per cent to 12 per cent for 1995, in response to falling EU stocks and below-average cereals harvests in Europe.

Criticisms of set-aside continued through the year, especially as newspapers began to realise how much money, as much as £1.25 million, could be earned by individual farmers, for 'doing nothing'. However, beneficial side-effects were seen to accrue, notably, a rise in land values of some 20 per cent and increased wildlife on set-aside land, and farmers were consistently advised by *Farmers Weekly* to 'use' their set-aside land in the most positive way possible. Nonetheless, the scheme remains a disgrace, and it is to be earnestly hoped that the post-1996 reforms will phase it out completely.

For more information on set-aside see the set-aside paragraphs in the Agriculture and the Environment section.

Elsewhere, the Potato Marketing Scheme was reprieved in November 1993 until 1997, and an August 1994 poll of producers voted overwhelmingly in favour of its extension after 1997. This vestige of the 1947 Act may be harder to kill than the government thought, especially since a shortage of potatoes increased prices threefold between October 1993 and 1994.

Table 3.2   *Arable area payment scheme 1994–5*

|  | Regional yield t/ha | Area aid £/ha | Set-aside £/ha |
|---|---|---|---|
| England | 5.89 | 247 | 313 |
| Wales non-LFA | 5.17 | 217 | 275 |
| Scotland non-LFA | 5.67 | 238 | 277 |

*Source*: MAFF news release, 5 September 1994

Finally, MAFF launched a three-part plan to develop alternative crops in July 1994, based on more research, the creation of a new 'Alternative Crops Unit' in the ministry, and a consultation booklet aimed at stimulating ways of integrating the work of all those concerned. Also in July, the ministry launched an Organic Aid Scheme (SI 1712/94) which offered payments over a five-year conversion period. Payments were set at £70 per hectare in the first two years, falling to £50 in the third year, £35 in the fourth year, and £25 in the fifth year. The scheme was part of the 1992 CAP Agri-Environment Package. Finally, a new Code of Good Agricultural Practice for the Protection of Soil was published in December 1993 as an ongoing commitment under the *This Common Inheritance* White Paper.

## Dairying

The main change during the year was the delayed agreement to phase out the Milk Marketing Board (MMB) as from 1 November 1994. Originally formed in 1933, the MMB's main role was to provide farmers with collective clout to protect against dairy companies pushing down prices. Over the intervening decades the MMB devised all sorts of schemes to encourage milk production in the most favourable areas, and the result has been a marked shift westwards in dairy production. Now, however, the MMB's successors, mainly Milk Marque (a voluntary producer-owned cooperative) but also food processors (for example, Nestlé), have been offering contracts to farmers, and the resulting milk will be auctioned to the dairy companies. First indications are that this will lead to a rise in the price of milk, and to a fall in manufacturing capacity as previously subsidised milk disappears from the market. The transition was the subject of an Agriculture Committee Report (HC 209(93–94)) which severely criticised the process and deeply regretted the frosty relations that had developed between the major supplier and the principal customers. The Committee promised to keep a close eye on the effectiveness of Milk Marque which set a price of 24p per litre for November 1994 in October 1994. In general, producers were receiving 2p per litre more for their milk after the changeover.

Changes were made to the dairy quota scheme in three SIs: 3234/93, 160-94, and 672/94 which provided a completely revised set of Regulations.

*Livestock subsidies*

There were several changes to the main schemes in this category: the Beef Special Premium; the Suckler Cow Premium; Hill Livestock Compensatory Allowances; and the Sheep Annual Premium. First, in January 1994 the ministry announced that payments under the Beef Special Premium Scheme would be nearly £90 million in 1993, compared to under £30 million in 1992, because the premium could now be claimed twice, and also because of a new extensification premium. For 1994, it was announced that the rate would rise from £56.34 to £69.07 for both the first and second claims. However, due to an overshoot of claims in 1993 some of the payments were scaled back in practice by around 25 per cent to around £42 in March 1994. Second, the rate for the Suckler Cow Premium rose from £65.73 in 1993 to £87.49 in 1994. Amendments to the scheme were made in SI 1528/94. Third, Hill Livestock Compensatory Allowances were cut by £25 million. In the Severely Disadvantaged Areas rates for cattle were cut by £15.80 to £47.50, and in the Disadvantaged Areas by £7.90 to £23.75. The cuts were occasioned by a rise in hill farmers' income, but, as the NFU pointed out, these incomes were still barely 60 per cent of the average wage. Changes to the Allowances Scheme were made in SIs 2631 and 2924/94 and 94/94. Fourth, the Sheep Annual Premium was raised by £1.76 to £19.24 for ewes.

In June 1994 limits to prevent overgrazing were imposed on the Suckler Cow Premium, the Sheep Annual Premium Scheme, and the Beef Special Premium Scheme, in order to bring them into line with limits already imposed on HLCAs. These were directed at a small minority of farmers who had been overgrazing, but the NFU expressed serious misgivings at the new measures. All these schemes were also subjected to quotas based on existing production in 1993, with 115 000 producers being given a quota. Amendments to the Regulations were made in SI 1528/94. Both these changes imposed considerable constraints on any expansion of hill farming, and hopefully will lead to better management of upland vegetation.

Finally, the Agriculture Committee carried out an investigation into the UK poultry industry (HC 67(93–94)), to which the government responded in HC 641(93–94). The report is interesting in that it describes the growth experience in an agricultural sector only slightly affected by the CAP. The industry accounts for 9 per cent of the UK's gross agricultural product, produces 27 per cent of the nation's meat, and provides some 45 000–50 000 jobs, mainly in slaughtering and processing. It is an industry in which a high degree of vertical integration between producers, processors and retailers has developed, and in which 10 firms produce 78 per cent of meat production, with two firms by themselves accounting for 34 per cent. The Committee expressed concern that this could lead to a loss of productive capacity to continental Europe, unless the government ensured a regulatory and economic climate which would favour UK production. In their response, the government replied that their general economic and deregulatory policies had produced the right conditions for UK entrepreneurs to flourish.

*The Integrated Administration and Control System (IACS) and summaries of grant schemes*

IACS is the system under which farmers register for payments under the main subsidy schemes: Arable Area Payments; Beef Special Premium; Suckler Cow Premium; Hill Livestock Compensatory Allowances; and Sheep Annual Premium. It suffered from a number of teething problems in its inaugural year (1993), but the existence of computer records and simplified procedures meant that the procedure went far more smoothly in the time leading up to the deadline for the second submission of forms in May 1994. Further amendments to the system were also set out in May 1994 in SI 1134/94.

Summaries of all the grant schemes are available in Nix, *Farm Management Pocketbook* (1994) and ATB Landbase, *Reaping the Benefits* (1994).

*Marketing*

A Marketing Development Scheme was launched in June 1994 (SI 1403 and 1404/94) as a replacement for the 1992 Group Marketing Grant. The scheme will last for three years, until 1997, with a budget of £10 million. Individual grants of 50 per cent up to a total of £150 000 are available under the scheme, which complements the £100 million available over 6 years under the FEOGA capital grant schemes for processing and marketing agricultural produce. Meanwhile, an Annual Report for the six remaining Agricultural Marketing Schemes was issued for 1991/2 by the Ministry in March 1994. Finally, the RSPCA launched a new farm animal welfare scheme, Freedom Food, which guaranteed consumers that pigmeat and eggs had been produced without cruelty.

*Tenancy reform*

In December 1993 the leaders of farm organisations reached agreement on proposals to reform the legislation regarding farm tenancies. Government proposals set out in October 1993 included: the creation of a new class of tenancy, the Farm Business Tenancy, which would actively encourage diversification; greater freedom for parties to negotiate their own tenancy agreements; and various safeguards. A Bill to implement the proposals was introduced in November 1994. Existing tenancies under the 1986 Agricultural Holdings Act will not be affected.

*Performance indicators*

The first two years of general set-aside produced differing harvest results. In 1993, the UK harvest at 19.5 million tonnes was down 9.5 per cent on 1992's 22.1 million tonnes. However, provisional figures for 1994 issued in October revealed a harvest of 19.9 million tonnes, up 2.5 per cent. In terms of crop area there was a 13.2 per cent fall in England and Wales between 1992 and 1993. This could not, of course, be repeated in 1994, but in the longer term the 1993 and 1994 crop areas of 2.55 million and 2.59 million hectares marked a dramatic fall from the 3.30 million hectares cropped in 1988. However, these reductions are constantly being countered by yield increases, with wheat yields, for example, rising from 6.70 to 7.30 tonnes per hectare in 1994.

More general indicators were provided by various MAFF and DAFS publications. For example: *Agriculture in the United Kingdom: 1993* (1994); *Agriculture in Scotland: Reports for 1992 and 1993* (Cm 2320 and 2607); and *The digest of agricultural census statistics: UK 1992* (1993) provide aggregate information about underlying trends, agricultural and environmental policy developments, and farm incomes. Farm incomes were the subject of *Farm incomes in the UK 1992/93* (1994) which revealed that total farm income across the UK had risen generally by 40 per cent between 1992 and 1993, while farm household income had risen by 62 per cent, thus reversing several years of general decline. To some degree the increases were due to the 1992 devaluation of the pound and the accompanying reduction in interest rates, which masked the long-term price reductions contained in the May 1992 reforms to the CAP. It may just be, however, that farmers are even more resilient than even the most optimistic would believe, and that Marxist theories that have predicted the end of family farming for decades may yet take a little longer to come true. Nonetheless, the long-term decline in the number of people employed in agriculture accelerated, and the labour force in agriculture fell at a rate of 1.6 per cent between 1986 and 1992, as compared to 1.2 per cent between 1979 and 1985, to a total of 598 000 in 1992 (Cm 2503).

*Concluding commentary*

In many ways the period between 1992 and 1996 must be seen as a transition stage, between one type of farm policy — the post-war expansionary model — and whatever may emerge after 1996. In the British context, further expansion would be useful, since we are only 73 per cent self-sufficient in the foods we can grow. However, we must follow the wider imperatives of Europe and the CAP. Evidence of continuing readjustment was provided during the year by several publications, for example: the OECD's, *Farm employment and economic adjustment in OECD countries* (1994); the Arkleton Trust's, *Final Report to the European Commission on Farm Household Adjustment in Western Europe 1987–1991* (1994); the Centre for Agriculture Strategy's, *The Recession and Farming: Crisis or Readjustment?* (1994); and the Centre for Rural Economy's, *Countryside Change: A synopsis of the ESRC Countryside Change Initiative 1989–1993* (1993). However, if readjustment is to make further progress it will need a more positive attitude from planners towards diversification than the one revealed by Percy, S. *et al.* in *Agricultural change and the planning response*, which found them unprepared to address the ramifications of agricultural change and what this meant for the redefinition of 'appropriate' economic uses in the countryside.

Against this background the NFU produced a 110-page document, entitled *Real Choices* in March 1994, which put forward four options for discussion. These were: Output Quotas; Input Quotas; Income Bonds; and Decoupling. The NFU President concluded that price cuts were inevitable. Books attacking the CAP and advocating a return to the free market continued to be published, for example, the graphically titled *Snouts in the Trough* by Atkin, M. (1994). Meanwhile, MPs and all the political parties campaigned for radical reforms of the CAP at the June 1994 elections to the European Parliament. ADAS in their *Agricultural Strategy* (1994) predicted that if current reforms continued agriculture would divide into two tiers: one tier based on large economic units and the other dependent on environmental schemes for their survival. Finally, a very important conceptual overview by Le Heron, *Globalised Agriculture* (1994), demonstrated very clearly how a global food regime has emerged, and why in a few years this may wield far more power, as post-war farm planning strategies are increasingly dismantled or modified, to take this into account.

## Agriculture and the environment

*Environmentally Sensitive Areas (ESAs)*

There were several changes to ESAs during the year: the designation of over 10 new areas; increased payments for the initial ESAs; and amendments to the boundaries of others. Taking each country in turn, in England six new ESAs were added to March 1994. These added over 300 000 ha to the scheme, bringing the total designated in all 22 English ESAs to 1 150 000 ha or some 10 per cent of agricultural land. Annual expenditure was planned to rise to £43 million by 1995/6.

In more detail, the six new ESAs were the Blackdown Hills (SI 707/94), the Cotswold Hills (708/94), Dartmoor (710/94), the Essex coast (711/94), the Shropshire Hills (709/94), and the Upper Thames Tributaries (712/94). Payments to farmers ranged from £12/ha to £310/ha. A new feature under the EC agri-Environment scheme was a payment of £710/ha for the provision of new opportunities for public access for walking and other quiet recreation.

Payments in the five first-round ESAs — the Broads (929/94); Pennine Dales (930/94); Somerset Levels (932/94); South Downs (931/94); and West Penwith (933/94) — were increased as a result of the two-yearly review by an average of 6.5 per cent in February 1994. For example, in the Broads, Tier 2 payments for grassland with an ecological interest rose from £220/ha to £225/ha, while Tier 3 payments for grassland with a high water table rose from £250/ha to £300/ha. New payments for access were also introduced for all ESAs as outlined in the previous paragraph.

Changes to other ESAs were made in SIs: 918/94 (North Kent Marshes); 919/94 (Test Valley); 920/94 (Suffolk River Valleys); 921/94 (Clun); 921/94 (North Peak); 923/94 (Breckland); 924/94 (South Wessex Downs); 925/94 (Lake District); 926/94 (South West Peak); 927/94 (Avon Valley); and 928/94 (Exmoor).

In Scotland, five new or improved ESAs covering the Cairngorm Straths; the Central Borders; the Stewartry; the Argyll Islands; and the Shetland Islands were designated (SIs 2345/93, 2767/93, 2768/93; 3136/93, and 3150/ 93). Extensions to the existing ESA in the Machair of the Uists and Benbecula, Barra and Vatersay were made in SI 3149/93. This brought the total area in the 10 Scottish ESAs to 491 285 ha, about 19 per cent of the land area.

In Wales, two new ESAs (the Clwydian Range and Preseli (SIs 238/94 and 239/94)) were designated, and modifications were made to the Yns Mon, Cambrian Mountains, and Lleyn Peninsula ESAs in SIs 2422/93, 240/94 and 241/94. This brought the total area in the seven Welsh ESAs to almost a quarter of the total land area.

The results of a major enquiry into ESAs were published during the year by academics from the Centre for Rural Economy at the University of Newcastle upon Tyne. They found that ESAs were cost-effective and a generally acceptable way of improving landscapes and wildlife habitat. However, they should perhaps be seen as stop-gaps en route to the possible long-term goal of ownership by amenity groups. The main findings are reported in Whitby, M. (ed.), *Incentives for Countryside Management: The Case of ESAs*, CAB International; Willis, K. from which Figure 3.1 is taken; Garrod, G. and Saunders, C., *Valuation of the South Downs and Somerset Levels ESAs*, The Centre; and in *Journal of Rural Studies*, 10, 1994, pp. 131–45.

**Figure 3.1**  *UK Environmentally Sensitive Areas 1994*

*Nitrate Sensitive Areas (NSAs) and Nitrate Vulnerable Zones (NZVs)*

There were several changes during the year. The main change was the inauguration of the process to implement the EC Nitrates Directive (91/676). This began with a consultation exercise in May 1994, which outlined plans for setting up 72 NVZs covering 650 000 ha. Within the proposed NVZs there would be compulsory controls on the application of both Chemical and Organic Nitrogen which would be legally enforceable, and field-by-field records would have to be kept for any allowable use. Detailed rules would be published by December 1995 with adoption at the farm level between 1996 and December 1999, the deadline set by the EC.

Reaction to the proposals was mixed. Arable farmers estimated they could carry on much as before, although livestock farmers were more concerned that some practices would have to be modified. All farmers, however, were worried by the lack of compensation in the NVZ scheme. Environmental groups, however, thought that the Zones were not extensive enough. Almost everybody expressed concern that the science itself was still to some degree uncertain, but work carried out by the Soil Survey showed that predictive models could be used on a field-by-field basis, to keep nitrate groundwater levels down to the 50 mg/litre set by the EC Nitrate Directive. However, the 50 mg/litre *maximum* limit for nitrate in drinking water was itself the subject of some doubt, with the NFU arguing that it was too low a limit, referring to a World Health organisation view that 50 mg/litre was acceptable as an *average* standard. During the consultation period which ended on 30 September 1994 the NRA called for the zones to be extended, while the CLA expressed dismay that the NVZs were no more than the first steps to greater restrictions.

Turning to NSAs, it was announced in December 1993 that payments to Premium NSA agreement holders would rise by £75/ha in 1994 in the 10 pilot NSAs launched in 1990 (SI 3198/93 and 3199/93), since research findings by MAFF also published in December — in *Pilot NSAs: Report on the First Three Years* — had shown that in nine out of the ten NSAs nitrate leaching had been reduced. More fundamental changes were announced in July 1994 with the designation of 22 new NSAs adding 35 000 ha to the 10 500 ha of the ten pilot NSAs (SI 1729/94). In these NSAs farmers can opt on a voluntary basis to make substantial changes to farming systems on a field-by-field basis, which would eventually be compensated for at a cost of £8.3 million a year. Individual payments would range from £65/ha/yr in the Basic Option B scheme to £550/ha/yr in some of the Premium Arable Option A areas, if land is both unfertilised and ungrazed. The NSAs will be subject to the same rules as the proposed NVZs (and indeed are contained within their catchments), but in most cases the NSA rules will be more demanding. The new NSAs were designated following consultation which began in August 1993 and had originally proposed 30 NSAs. Only two of these had been abandoned and the remaining six had been amalgamated with others to form the new 22 NSAs. Two NSAs were proposed for Scotland in June 1994. Finally, in October 1994, MAFF announced that the original ten NSAs would be extended for a further five years, with terms and conditions similar to the 22 new ESAs announced in July. Some 160 farmers had participated in the scheme covering some 87 per cent of the agricultural land in the NSAs.

Elsewhere, in August 1994, MAFF announced intensive campaigns in seven areas encouraging farmers to prepare individual management plans for spreading manure and similar organic wastes in order to cut river pollution. Earlier, in May 1994, the NRA had produced a list of 33 inland waters thought to be sensitive due to their eutrophic status. The NRA also produced a list of areas where phosphorus-removal plants would be required by the end of 1988 in order to reduce the nutrients causing eutrophication.

## The Farm and Conservation Grant Scheme

This scheme was set up in February 1989 to help farmers maintain efficient farming systems while also combating pollution and conserving the country-side and its wildlife. Under the scheme farmers can apply either for a part EC-funded Improvement Plan or a nationally funded grant on one-off investments. Between February 1989 and November 1993 £149 million was paid to over 9000 farmers under the scheme. Most of the funding, £119 million, was spent on installing or improving waste-handling facilities. In November 1993 MAFF announced that funding for the scheme would be reduced by £7 million in 1994/5, and that rates of grant would be cut. For example, grants for creating or improving hedges were reduced from 50 per cent in the LFAs and 40 per cent elsewhere to 35 per cent and 25 per cent. However, new grants of 30 per cent in the LFAs and 25 per cent elsewhere were introduced to help farmers who wish to facilitate public access to set-aside land meet the cost of signposts and stiles. These and other changes to the scheme were set out in SIs 2900/93, 2901/93 and 1302/94.

## New Habitat Scheme

A new environmental scheme to encourage farmers to create and enhance important wildlife habitats, by taking carefully selected areas of land out of production for 20 years, was launched by MAFF in May 1994, with a planned expenditure of £3 million per year. Three different options under the generic title of the Habitat Scheme were offered: first, the establishment or enhancement of *water fringe habitats* in six designated pilot areas, for example, the Upper Avon in Wiltshire (SI 1291/94); second, management of wildlife habitats which have been established under the Five-Year Set-Aside scheme (SI 1292/94); and third, establishment of areas of *saltmarsh* on suitable coastal land (SI 1293/94). Annual payments per hectare ranged from £125 to £360 in the water fringe scheme from £195 to £525 in the saltmarsh scheme, and a standard payment of £275 in the *former set-aside land* scheme. The Habitat Scheme is one of the six agri-environment schemes which is assisted by the EC under the May 1992 reforms to the CAP. Although the schemes were generally welcomed, the RTPI criticised the increasing number of schemes on offer, their general lack of coherence, and failure to offer an integrated solution.

### Stewardship and Hedgerow Schemes and hedgerows in general

In June 1994 MAFF announced that responsibility for the Countryside Stewardship and Hedgerow Incentive Schemes would be transferred from the Countryside Commission to MAFF on 31 March 1996. The necessary legislation would also confer a statutory advisory role on the DOE, the Countryside Commission, English Nature and English Heritage on not only the future of these schemes but on all other MAFF environmental land management schemes. The government also promised to examine whether all the environmental schemes could be better integrated. The Countryside Stewardship Scheme was launched in June 1991 as an experimental scheme in five key landscape types, to protect, enhance and regenerate them, as well as improving public enjoyment. Two more landscape types were added in 1992. The 1994/5 budget for both schemes was £13 million, and between 1991 and 1994, 3894 agreements (931 of which were hedgerow agreements) had been made, covering 80 000 ha. About 13 000 ha of open access and over 300 km of linear access had been provided by the agreements.

Although the schemes have shown what can be done, and have been generally welcomed, they are in effect only drops in the ocean. This was demonstrated by a DOE *Hedgerow Survey 1990–93* published in July 1994 which showed that hedgerow loss was still continuing, albeit at a rate of 3600 km per year, compared to 9500 km per year in 1984–90. More encouragingly, the survey also revealed that the rate of removal is not outweighted by the rate of new planting which had risen from 1900 km per year in 1984–90 to 4400 km per year in 1990–93. However, the rate at which hedges were becoming derelict through poor management had increased from 7400 km per year to 22 500 km per year. The CPRE argued that this pattern of loss and dereliction was unacceptable and highlighted the need for increased resources for the Hedgerow Incentive Scheme.

The Hedgerows Bill (HC 31(93–94)) was the latest in a long line of attempts to introduce planning controls over hedgerows (for example, see Volume Four, p. 92). This Bill, which only received its first reading, would have made it an offence to uproot or remove a hedge which either

(1) followed a parish boundary, or the boundary of a landholding, or which bounds any footpath, bridleway, byway or public road; or (2) had been deemed to be of significance in the landscape for environmental reasons by the local planning authority. Such hedges could have been removed if planning permission was granted, as a result of an emergency, or if their removal was necessary for proper maintenance or for access. The Bill would also have given powers to restore hedgerows. Measures to safeguard hedgerows were, however, signalled in the Queen's Speech in November 1994.

*Environmental aspects of set-aside*

In September 1994 MAFF launched the *Countryside Access Scheme* with an annual budget of £4.5 million. Under the voluntary scheme farmers can receive annual payments of £90 per ha for access routes and £45 per ha for open field sites, in return for managing their set-aside land for public access under a five-year agreement.

However, elsewhere the CPRE claimed that set-aside was producing a minimal environmental dividend, and that support for the environment was a small fraction of the amount spent on supporting farmers and farming, as shown in Table 3.1. In a similar vein, the Countryside Commission issued a statement on set-aside (CCP441) which argued that set-aside should not be used as the dominant means of reducing crop production, and that schemes that offered environmental benefits as the primary purpose and reduced crop production as a side-effect should be encouraged instead. Where public money continues to support set-aside it should deliver real public benefits in terms of the appearance, wildlife and public enjoyment of the countryside. This statement was based on the findings of a report on the *Countryside Premium for Set-Aside Land* (CPP 428) which revealed that set-aside land could deliver significant environmental benefits if incentives are available, but that set-aside is not the ideal starting point for an environmental scheme. Finally, the Institute of Terrestrial Ecology produced a guide for *Managing set-aside land for wildlife*.

*Commentary*

The year thus saw the launch of five of the six EC-assisted agri-environmental reforms developed as a consequence of the May 1992 reforms to the CAP, namely: new ESAs; the Habitat Scheme; the extended NSA scheme; the Organic Aid Scheme; and the Countryside Access Scheme. MAFF predicted that the sixth scheme, the Moorland Scheme, would be launched by the end of 1994. These new schemes added to the already overlong list of schemes available which were listed by *Town and Country Planning* in the April 1994 edition on p. 120 as follows: Farm and Conservation Grant Scheme, Farm Woodland Premium Scheme, Nature Conservation Grants, Landscape Conservation Grants, Woodland Grant Scheme, Ancient Monument Grants, Countryside Schemes, Wildlife Enhancement Scheme, ESAs, Countryside Stewardship, NSAs, Hedgerow Incentive Scheme, Hedgerow Renovation Scheme, Management and Nature Reserve Agreements, National Parks Agreements, Local Authority Agreements, Ancient Monuments Agreements, Inheritance Tax, and Access Agreements.

# Forestry

*Review of Forestry Commission and grant schemes*

After a year long-review the government decided not to privatise the Forestry Commission in Cm 2644, *Our forests: The way ahead: Enterprise, Environment and Access: Conclusions from the Forestry Review.* Instead, Forest Enterprise will become a Next-Steps Agency (Trading Body) run by a chief executive. Forests owned by Forest Enterprise may still be sold, particularly those with little benefit for access, wildlife, or landscape, but those where there is a high level of access will be made more difficult to sell. An extra £1 million will be given to the Forestry Commission to buy out leases where access is restricted. The target for afforestation remained unchanged at 33 000 ha/yr with Scotland the main target being zoned into planting areas. Turning to incentives, no new tax reliefs were announced but other incentives were as follows (Cm 2645):

(1) Conifer planting grant up from £615 to £700/ha
(2) Broadleaved planting grant rate £1350 below 10 ha (formerly £1175–1575), and £1050/ha in areas above 10 ha (formerly £975) (both (1) and (2) paid in two instalments, 70 per cent after planting, 30 per cent after 5 years
(3) Better land supplement for conifer and broadleave planting up from £400/ha to £600/ha
(4) Restocking grants of £325/ha and £535/ha for conifers and broadleaves respectively to be paid in a single instalment instead of over 10 years
(5) Natural regeneration, a discretionary grant of 50 per cent of the approved cost. After regeneration further grants of £325/ha for conifers and £535 for broadleaves
(6) Premium payment for exclusion of livestock from old-established or native woodlands threatened by grazing of £80/ha/annum for 10 years
(7) £1 million a year for pilot schemes in priority target areas
(8) A new discretionary Woodland Improvement Grant to improve environment and visitor facilities

(9) New grants for short-rotation coppice at £400/ha on set-aside land and £600 elsewhere to run for 5 years initially

(10) New annual management grant of £35/ha to replace the standard management grant

(11) Pilot scheme under which grant payments for larger Woodland Grant Scheme applications will be determined by negotiation rather than by reference to specific rates of grant.

Other sources of aid under the Farm and Conservation Grant Scheme and the Farm Woodland Premium continued unchanged.

Altogether the initiatives would add £4 million to forestry spending, but were subject to EU approval which was sought in September 1994. The proposals received a generally favourable response but the CPRE warned that the multipurpose objectives may be crowded out by commercial considerations, and that the continued (albeit slow) rate of forest sales amounted to privatisation by stealth.

*Forestry policy*

The main aims of Forestry Policy as set out in the 1991 statement — the sustainable management of existing woods and forests and a steady expansion of tree cover to increase the many diverse benefits that forestry provide — were incrementally added to by Cm 2429, *Sustainable Forestry; the UK programme*, which incorporated further objectives arising from commitments made at Rio in 1992. These policies were set out in typical prose accompanied by lavish colour illustrations under eight headings: Protecting our forest resources; enhancing the economic value of our forest resources; conserving and enhancing biodiversity; conserving and enhancing the physical environment; developing the opportunities for recreational enjoyment; conserving and enhancing our landscape and cultural heritage; promoting appropriate management; and promoting public understanding and participation. The document also set out a framework for expanding the woodland and forest cover. The modified policies are certainly to be welcomed, but they are incremental rather than revolutionary and forestry policy will still have vehement critics, although it has made great strides towards a genuinely holistic approach in recent years.

*Tree Preservation Orders*

In July 1994 the DOE announced the outcome of its review of TPOs which it had set up in 1990. This found support for the system but with some respondents calling for measures to encourage the proper management of trees. This was rejected, but the government announced that it planned to change the law to make it more difficult to damage or destroy trees by amending the law relating to 'wilful' damage to 'reckless' damage, and by making it harder to cut down dying trees.

## Specific reports and reviews

The expenditure plans for the Forestry Commission between 1994–5 and 1996–7 were set out in Cm 2514 which revealed that its subsidy would remain at around £94 million per year over the period, but that total grant payments would rise to £34.1 million by 1996–7. Provision was made for 54 884 ha of new planting, of which 52 560 ha would be carried out by the private sector over the three years, and for 47 000 ha of regeneration by the Commission.

Previous expenditure and work by the Commission was examined in the *73rd annual report and accounts for the year ended 31 March 1993* (HC 151(93–94)), which revealed that the Commission grant aided 15 295 ha of new planting in 1993 and planted 2356 ha of new planting itself. Total new planting and restocking in both sectors totalled 34 199 ha. More detailed aspects of the Commission's work was provided in *Report on forest research for the year ended March 1993*.

In Wales, a Report by the Welsh Affairs Committee on *Forestry and Woodlands* (HC 35(93–94) made an amazing 45 recommendations which basically called for a stable regime aimed at a steady expansion of forestry that would be both environmentally and commercially attractive, and advocated a planting rate of 2400 ha per annum, which would provide a 50 per cent increase in forestry by the year 2050. In their reply, the government (Cm 2645) did not feel it appropriate to set a target but reaffirmed its commitment to an expansion in the tree cover and pointed to the better targeted grants which were available following the Forestry Review.

Academic research by Watkins, C. *et al., Farmers not foresters*, found that few farmers are tempted to become foresters by current grants. The reasons farmers do not plant woodland were found to be: inadequate financial incentive (87 per cent); long pay-back time (47 per cent); loss of flexibility (40 per cent); public access concerns (33 per cent); and uncertainty (20 per cent). In contrast, research for the Forestry Commission, *The influence of woods in property values*, found that woodlands could increase values by at least 10 per cent, and the CPRE revealed the results of a survey which showed the great value placed by the public on woods.

## Specific forests

In July 1994 the DOE announced that the 200-square-mile National Forest would be run as from April 1995 by a specially set-up independent public company, who would take over the pioneering work of the Countryside Commission. The target of the new body would be to create new woodlands over about one-third of the area. Elsewhere, the DOE announced its response to a 1992 consultation paper on the future of the New Forest. This rejected the original idea of giving the area a statutory designation, but instead announced that the area would be treated as if it were a national park with the GDO being amended accordingly in order to extend stricter development control to the area. The New Forest Committee were retained but again were not given statutory status as proposed.

*Advice*

The Forestry Commission published a series of advisory documents including: *Forest Landscape Design Guidelines* and *Forests and Water: Guidelines* (third edition) both of which were models of visual presentation. The Commission also published *Creating new native Woodlands, Reclaiming disturbed land for forestry*, and *The Wildlife Rangers Handbook*.

# Water and the coast

*The Land Drainage Act 1994 (Chapter 25)*

This amended the functions of internal drainage boards and local authorities by inserting after Section 61 of the Land Drainage Act 1991 a set of duties relating to conservation of the natural and built environment and the provision of recreational access.

*The National Rivers Authority (NRA)*

The NRA continued to grow in stature with a series of publications. The most significant was *Water: nature's precious resource: an environmentally sustainable water resources development strategy*. This emphasised the need to control demand by selective domestic metering, and to reduce leakages in the distribution system, so that large-scale infrastructure schemes, such as new reservoirs and water transfer schemes, can be avoided in the next 20 years. The alternative could be an average rise in demand of 25 per cent (and 43 per cent in the south of England), leading to acute water shortages, near-permanent hosepipe bans, threats to public health, and pressure on rivers, lakes and wildlife. In the medium term the NRA's *Corporate Strategy* set out three key areas: holistic integrated environmental management; strategic planning; and evaluating success against measures such as water quality. In the short term the *Annual Review 1993–94* and its *Accounts for 1991–92 and 1992–93* (HC 83 and 84(93–94)) recorded that 32 000 pollution incidents had been dealt with, and the *Water quality in rivers 1990–92 survey* noted that 11 per cent of the total length of waterways had been upgraded. The NRA also published *Discharge Consents and Compliance; Water for farm irrigation* which noted that irrigation was in the national interest and was a necessity rather than a luxury; *Guidance from the NRA to local authority planners* on how to treat water issues in their work; *River landscape assessment: methods and procedures; River corridor surveys: methods and procedures*; and *Development of environmental economics for the NRA*.

*Water provision and quality*

In August 1994 the DOE received a report on the system by which grants have been paid towards first-time connection to mains water and sewerage services for 50 years. Almost a million households, however, still lack mains sewerage. Accordingly, the DOE issued a consultation paper on the future of the scheme and attitudes to imposing sewerage controls on new development. This proposed restricting the connection grant to where there was a sound environmental and financial case, and making occupation of a new dwelling conditional on adequate sewerage arrangements. In July 1994 the water companies were given permission to raise their charges above inflation in order to improve water quality by investing in better infrastructure.

*Advocacy*

The CPRE published *Water for Life*, which called on government at all levels to integrate water issues in forward planning and development control. English Nature, in *Water Quality for wildlife in rivers*, highlighted the growing problem of excessive phosphates in at least 20 rivers and called for £9 million to be sent on removal. Finally, Pond Conservation in *A future for Britain's ponds* outlined an agenda to halt the decline estimated to 75 per cent over the last century.

*The coast*

The House of Commons Environment Committee in HC 294(93–94) took evidence on *Coastal zone protection and planning* from the DOE and MAFF which related to two consultation papers issued by the DOE in October 1993 on *Development Below the Low Water Mark* and *Managing the Coast*. The Committee expressed regret that the severe weaknesses they had identified in a previous report on the topic had been largely ignored, but in July 1994 the DOE decided to make no change to the existing statutory systems for coastal management, in spite of widespread comments that it was unwieldy, incoherent and prone to misunderstandings. The consultation was informed by a consultant's review for the DOE entitled *Coastal planning and management*. Also in October 1993 MAFF announced a new policy on coastal defence which emphasised 'soft defences' based on encouraging natural barriers like saltmarshes. Recent achievements by local authorities were outlined in King, G. and Bridge, L., *Directory of Coastal Planning and Management Initiatives*, and Andrew, D. and Pinney, D. (eds), *Coastal Planning Management*.

## Conservation and recreation

*Organisational structures and Statement of Intent*

The main story of the year was the proposed merger of the Countryside Commission and English Nature following the success of the same kind of merger in Scotland and Wales in the early 1990s. The process began with the issue of a DOE consultation paper in February 1994 which asked the 400 consultees if there should be a merger, and if so what the remit of the new body should be. In a useful statement and accompanying briefing notes the Countryside Commission pointed out the risks involved, and argued that a merger should only be agreed to if these risks could be conclusively removed. In contrast, the Council of English Nature expressed enthusiasm for the project. They were supported by the Association of National Parks. However, almost every other group expressed doubts or reservations: for example, the CPRE, the Town and Country Planning Association, the Association of County Councils, the RTPI, the Ramblers Association, and the House of Lords. While none of these groups were against the merger in principle, in practice they feared that more would be lost than gained. Commentaries on the proposals were made in: *Town and Country Planning*, May 1994; *Ecos*, 14(3/4) 1993; and in Grove-White, R., *England's Green Horizon? A conservation and countryside access agency for the post-Rio world*. The sheer weight of opposition to the merger convinced ministers in October 1994 that the merger should be abandoned, but that there should be closer cooperation between the two bodies.

The other main story of the year was gradual progress towards setting up Environment Agencies for England and Wales, and for Scotland. Paving legislation which would allow expenditure to begin and for shadow bodies to operate was announced in the Queen's speech in November 1993. This was followed by a consultant's report by Touche Ross on *Options for the Geographical and Managerial Structure of the Proposed Environment*

*Agency*, which was published by the DOE in July 1994. Draft legislation was issued by the DOE in October 1994 which envisaged that the Environment Agency would be formed from a merger of the National Rivers Authority, Her Majesty's Inspectorate of Pollution, and local government waste regulation departments. Environmental groups criticised the draft as being far too weak, although a former critic, the NRA, rallied behind the key principle of integrated environmental management, covering water, land and air. A Bill to set up the Agencies was announced in the Queen's Speech in November 1994.

Apart from these potentially major changes the Countryside Commission published advice on how voluntary organisations could achieve even more as a key component in achieving a sustainable countryside, and how the *Countryside Commission and Voluntary Organisations* (CCP460) could work even more closely, in addition to the annual funding of £2.5 million which the Commission already provides for the voluntary sector. For example, the CPRE in their *1994 Annual Report* demonstrated just what an important job the voluntary sector performs. Another major contribution made by the Commission is by assembling information on *Training for Countryside Managers, Staff and Volunteers* (CCP 372).

The year also saw a clutch of annual reports and the now obligatory corporate plans. For example, the Countryside Commission published a summary of its corporate plan for 1994/5 to 1997/8 in CCP 467, its Annual Report for 1993–4 in CCP 456, and its accounts for 1993–4 in HC 632(93–94). Accounts for the Countryside Council for Wales were published in HC 919(92–93). English Nature decided to restructure its internal organisation as from April 1994 in line with its *Strategy for Nature Conservation in the 1990s* and its concept of 'Natural Areas'. English Nature also published its *2nd Report 1992–93*, which contains a detailed description of its five-fold work plan; *Progress '94*, which provides a more popular and wide-ranging review of English Nature's advocacy role, under the slogan of 'maintain, sustain, campaign'; and its *Research Programme 1993/94*. The final accounts of English Nature's predecessor were published in HC 626(93–94).

Elsewhere, Scottish Natural Heritage published its *Review of our First Year*, as a summary of its first *Annual Report for 1992–93*; its *Second Operational Plan: Review of 1992–93 and Work Programme 1993–94*; its *Accounts 1992–93* (HC 177(93–94) and also a *Charter Standard Statement* on how it would incorporate the Citizens' Charter into its work. These demonstrated how much England will lose by not merging the Countryside Commission and English Nature, along the model pioneered by Scottish Natural Heritage.

Finally, English Nature signed important statements of intent with the National Parks and the Association of County Councils. The statement commits the signatories to give nature conservation a high priority in key decision-making processes, to develop nature-conservation strategies, and to integrate nature-conservation objectives with all aspects of plans, policies and proposals.

## National Parks Bill 1994 and AONBs

Another abortive change during the year was the attempt to legislate for independent national park authorities, with sole powers over planning and minerals matters, as recommended by the 1991 report of the National Parks Review Panel. Although the government had accepted the principle, they failed to find parliamentary time for the legislation, and so a Private Member Bill was introduced into the House of Lords in March 1994 (HL 32(93–94). The Bill was also introduced in order to forestall any changes that local government reform might bring about (see pp. 87–88). Although the Bill received widespread and all-party support, and completed its passage through the Lords, it ran out of time in May 1994 when a former Minister of Agriculture, Michael Jopling, objected to a lack of time for debate during its first reading in the Commons. However, the government promised a new Bill which would legislate not just for independent national park authorities but also for those other recommendations made in the 1991 report, which had been accepted by the government not only in their response to the report but also in their 1992 election manifesto. A Bill was announced in the Queen's Speech in November 1994.

Areas of Outstanding Beauty received a boost with the publication of *A Guide for Members of Joint Advisory Committees* in CCP 461 which was accompanied by an updated publicity leaflet CCP 276. The Guide stressed the need for management plans based on partnership, and the appointment of an AONB officer at a senior level. However, *Planning* (23 September 1994) reported that only two-thirds of AONBs have the advantage of a joint advisory committee, and only one-quarter have appointed a specialist AONB officer. In February 1994 the 40th AONB was designated, Nidderdale in North Yorkshire.

## Nature Conservation

*Legal cases*   For the first time since it was set up in 1991, English Nature used its powers to prosecute over damage caused to an SSSI, when it successfully prosecuted Easthome Properties who were fined £9140 in March 1994 after they had been found guilty of 14 offences in the South Thames Estuary and Marshes SSSI. However, in August 1994, the RSPB lost a legal challenge in the Court of Appeal over the DOE's decision in December 1993 to exclude the 22 ha of Lappel Bank from the designation of the Medway

Estuary and Marshes as a Special Protection Area (SPA), on economic grounds, since it could be needed for a possible expansion at the port of Sheerness. The decision was thought to be an important test case, since although it was fought under the terms of the 1979 Birds Directive, it was thought that it would also colour reactions to the consultation paper on the implementation of the Habitats Directive which was issued at the same time. The key point of concern was the degree to which economic grounds could or could not be used to overrule conservation arguments.

*Birds Directive*   French proposals to change the 1979 Birds Directive were considered by the Select Committee on the European Communities in *The Protection of Wild Birds* (HL 70(93–94)). This concerned a proposal made at the March 1994 meeting of the European Council to alter the closed season for the hunting of birds in order for individual countries, and regions within countries, to bring the closed season into line with what was actually happening on the ground, i.e. hunting outside the closed season. These breaches had already been acknowledged by the Commission in their November 1993 report on the implementation of the Directive (COM(93) 572 final). The Committee concluded, however, that the proposals would make effective enforcement even more difficult than at present.

*International agreements and designations*   In July 1994 the DOE welcomed the adoption of a new strategy for conserving migratory species worldwide under the Bonn Convention on Migratory Species. New designations included: in November 1993 Salisbury Plain as an SPA; in December 1993 the Bowland Fells in Lancashire, and Stodmarsh** in Kent as SPAs; in February 1994 the Benfleet and Southend Marshes** in Essex, and Thursley Hankley and Fensham Commons* in Surrey as SPAs; in July 1994 the Stour and Orwell Estuaries** in Suffolk as SPAs, but with exclusions to allow the expansion of the Port of Felixstowe; and in September 1994 Broadland** as an SPA. Some of these sites (marked with an ** or *) were also entirely** or partly* designated as Wetlands of International Importance under the Ramsar Convention. Other sites to be listed as Ramsar sites during the year included Malham Tarn in the Yorkshire Dales and the Midland Meres and Mosses in Cheshire. These designations brought the UK total of SPAs to 96 and Ramsar Sites to 83 by September 1994. Nationally, Slapton Ley in Devon and Bassenthwaite Lake in the Lake District were made National Nature Reserves (NNRs). In Scotland, five SPAs were designated in April by Scottish Natural Heritage, four of them in Orkney and Shetland, the first to be made in that area, and the fifth was in Angus, the Loch of Kinnordy. Also in Scotland the RSPB acquired 220 hectares in the Cairngorm Straths ESA. In Wales, the Countryside Council for Wales declared two more NNRs to bring the total to 52 by July 1994.

*National Nature Reserves and Sites of Special Scientific Interest* The National Audit Office conducted an investigation into the methods of *Protecting and managing sites of special scientific interest* (HC 379(93–94) which also covered NNRs, since NNRs are also SSSIs. This found that the 3700 SSSIs and 140 NNRs were being well managed by English Nature, and that though notification by itself does not guarantee protection it does provide strong safeguards. The Office also found that between 90 per cent and 95 per cent of the potential sites had been designated. Deficiencies were, however, found in monitoring procedures, with some habitat surveys out of date. Over 850 sites had experienced some damage between 1987 and 1994, albeit mostly of a short-term nature. More work also needs to be done on approving management plans, since, to date, only half of the sites have approved plans, with a further third having plans either in draft or awaiting approval. With around 3 million visitors a year the Office also argued that the public should be given more information and better access, but that this should be concentrated on those reserves that provide the best access, interpretation, and involvement for the public. Conspicuously, the report did not criticise the legislation behind the processes it examined, but did observe that 'English Nature's position is strongest when they own and manage the land which is the case in nearly one quarter by area of reserve land' (p. 28). Finally, although expenditure on management agreements has risen from £0.3 million in 1983–4 to £7 million in 1992–3, recent payments have concentrated on shorter-term positive agreements or the Wildlife Enhancement Scheme, and Compensatory agreements are now only being offered in exceptional circumstances. In spite of this, the DOE announced a further £800 000 for management agreements in SSSIs in February 1994. However, two other reports (English Nature, *SSSIs in England at risk from acid rain*, and Countryside Council for Wales, *Terrestrial SSSIs at risk from soil acidification in Wales*) attempted to show how many SSSIs may be at risk from acid rain. In both cases the risks vary with the amount of acid rain predicted under different scenarios, and using different spatial scales. These scenarios are also discussed by Farmer, A. and Bareham, S. in *The Environmental Implications of UK Sulphur Emission Policy Options for England and Wales.*

*Implementation of the EC Habitats Directive* The process began in October 1993 when the DOE published a consultation paper on the implementation of the Directive (92/43) on the Conservation of Natural Habitats and of Wild Fauna and Flora. This was followed by draft regulations in July 1994. These proposed new provisions for site selection and designation, so that existing SPAs can contribute — along with the new designation of Special Areas of Conservation (SACs) — towards the UK contribution to the European Union network of designated areas to be known as Natura 2000. The draft regulations also proposed a number of changes to existing legislation and duties to take account of the extra designation of SACs, notably reinforcement of the Town and Country Planning and highways legislation to ensure that harmful developments would not be allowed in either SPAs or SACs, which in some cases could lead to revocations of existing permissions. The draft included a map of potential SPAs and flow charts for considering development proposals in SPAs and SACs. The draft regulations were welcomed by English Nature, but the RSPB complained that instead of introducing primary legislation the government had simply amended existing legislation, thus missing a golden opportunity to improve wildlife conservation, rather than adding another layer of designation and thus confusion and complexity. The regulations also continued the practice of ambiguous phrases which would allow development when the overriding public interest was greater than the nature conservation interest, thus keeping the door open to development. Further doubt was cast on the government's commitment when English Nature submitted its draft list of SACs to the government, all of which were either SSSIs, NNRs, or SPAs already, and observed that in most cases, listing as a SAC would make little or no difference to existing management. In conclusion, a report by Derek Ratcliffe for Friends of the Earth, *Conservation in Europe, will Britain make the grade?*, summed up the general mood of disillusionment with a government that once again had failed to meet any more than the letter rather than the spirit of European law.

*Peatland*   In September 1994 new measures to protect peatlands were issued in a draft Minerals Planning Guidance Note which advised that any future peat extraction should be restricted to areas which have already been damaged, and are of limited nature conservation value. However, although the draft Note also proposed updates of existing permissions, environmental groups wanted a more aggressive approach which would lead to extraction being prevented on valuable sites as soon as possible, and a greater use of alternatives. These alternatives were considered by the RSPB in *Growing Wiser* and by the DOE *Report of the Working Group on Peat Extraction and Related Matters* whose work had preceded the draft Note. Meanwhile, in July 1994 English Nature and Fisons reached an agreement under which Fisons donated 3240 ha of peatland with permission for extraction to English Nature for long term conservation. Under the deal some extraction will still continue but in a less damaging way.

*Publications*

English Nature marked its new presence with a flurry of publications which included: *Nature Conservation Strategies — The Way Forward*, a guide for local authorities contemplating the production of a strategy; *England's*

*National Nature Reserves*, a popular guide to the Reserves; *Managing Local Nature Reserves*, guidance to local authorities on how to combine conservation with educational, public, and scientific use of a reserve; *Nature Conservation in environmental assessment*, a guide for planners and developers; *Roads and nature conservation — guidance on impacts, mitigation and enhancement*, a good practice guide on how to minimise the effects of road building; *Nature conservation guidelines for renewable energy projects*, an examination of the issues and possible side-effects, including more income for agricultural improvements, which might damage wildlife; and *Planning for wildlife in towns and cities*. This last document was complemented by CCP 462, *Securing a greener future for London*. In Scotland, Scottish Natural Heritage issued *Red Deer and the Natural Heritage: a policy paper*, which proposed a reduction of 100 000 animals, from the current total of 300 000, as a first step to integrated and sustainable management, and the Cairngorms Working Party produced recommendations for an integrated management strategy for the Cairngorms to be run by a single special authority in *Common Sense and sustainability: a partnership for the Cairngorms*.

The County Planning Officers Society in *Caring for Nature* reviewed existing progress in local authorities and set out a six-stage approach towards good practice including identifying the resource, evaluating it, preparing a strategy, preparing a management plan, monitoring and review, and consulting and encouraging participation. Existing work by local authorities is described by the RSPB in *Strategies for Wildlife*, which reveals that about a third have a strategy already, and that about 80 per cent of the remainder plan to do so. The work of the Local Government Nature Conservation Initiative in promoting a non-statutory system of 'Sites of Importance for Nature Conservation' is described by Collis, I. and Tyldesley, D. in *Natural Assets*. More prosaically, Reid provides an up-to-date text on *Nature Conservation Law*, Friends of the Earth provide splendid information for campaigners in *A Citizen's Guide to Environmental Rights and Action in Scotland*; Poyser T. and Poyser A.D. provide empirical evidence of a decline in 24 out of the 28 species of farmland birds, but falls are matched by increases in the 174 non-farmland birds in *The New Atlas of Breeding Birds: 1988–91*; other losses are recorded by the RSPB in *Wet Grasslands — What Future? An account of wet grassland loss in the UK*; but the World Wide Fund for Nature in *The effectiveness of Riparian Buffer Strips for the control of agricultural pollution* show how these relatively small pieces of land, now often put into set-aside, can provide disproportionate gains.

One of the central messages of conservation is the need to take an integrated and holistic viewpoint. However, Redclift, M. and Benton, M. (eds), in *Social Theory and the Global Environment*, argue that the scientific community has hijacked the environmental debate and has restricted the agenda to scientific solutions, thus precluding discussion about the real cause, humanity's exploitative attitude to the environment. Human attitudes to the environment are covered in: Milton, K. (ed.), *Environmentalism: The View from Anthropology*; Matthews, F., *The Ecological Self*; Simmons, I.G., *Interpreting Nature: Cultural Constructions of the Environment*; and Smil, V., *Global Ecology: Environmental Change and Social Flexibility*. Human attitudes as expressed by politicians and diplomats are explored in O'Neill, J., *Ecology, Policy and Politics: Human Well-Being and the Natural World*; Suuskind, L.E., *Environmental Diplomacy, Negotiating more effective global agreements*; and Young, S.C., *The Politics of the Environment*. The net results of political decisions are revealed by the New Economics Foundation in *Green League of Nations*, which ranks the 21 richest nations by their environmental record (the UK came tenth). Other issues are covered in Wall, D. (ed.) *Green History: A reader in Environmental Literature Philosophy and Politics*; Interdisciplinary Research Network on Environment and Society, *Perspectives on the Environment*; Berry, R.J. (ed.), *Environmental Dilemmas, Ethnics and Decisions*; and Conford, P. (ed.), *A Future for the Land*, which examines organic agriculture and forestry. Finally, Pickering, K.T. and Owen, L.A., in *An Introduction to Global Environmental Issues*, provide a well-illustrated account which combines scientific and attitudinal data quite well, but still fails to redress the scientific imbalance described at the beginning of this paragraph.

O'Riordan, T. (ed.) provides a technically orientated text on *Environmental Science for Environmental Management* but fails to show how these can be translated into real action on the ground. For example, there is virtually no discussion of planning for any form of land use, and planning does not even merit an index entry. Ways in which to use environmental information are also examined in Michener, W.K. *et al., Environmental Information Management and Analysis: Ecosystem to Global Scales*; and McCloy, K., *Resource Management Information Systems*. The particular case of *Impact Assessment and Sustainable Resource Management* is considered by Smith, L.G. while economic methods are discussed by Tietenberg, T., *Economics and Environmental Policy*, and Hanley, N. and Splash, C.L., *Cost-Benefit Analysis and the Environment*.

Other methods for achieving conservation goals are set out in CCP437, *Using covenants for conservation*; Rural Action in *Improving the Environment*, which advises parish councils on how to use statutory powers; Woollett, S., *Environmental Grants: A guide to grants for the environment from government, companies and charitable trusts*; and Wilson, C. (ed.), *Earth Heritage Conservation*, which provides practical guidance for safeguarding the geological heritage. Legal issues are covered in Polden, M. and Jackson, S., *The Environment and the Law: A Practical Handbook*, and Gregory, M., *Conservation Law in the Countryside*. Finally, Goldsmith, F. and Warren, A., *Conservation in Progress*, is the third book to spring from the well-respected course in conservation run by UCL, while *The findings of the People,*

*Economics, and Nature Conservation research programme funded by ESRC/
NCC* are edited by Burgess, J. from the UCL Biology Department.

*Landscape and other surveys*

The main event was the publication of the *Countryside Survey 1990* of the
UK by the DOE based on work by the Institute of Terrestrial Ecology. This
compared data from 1990 with data from 1978 and 1994, using a
combination of field survey of 508 sample 1 km squares and satellite
imagery. Although not a comprehensive record of change it revealed a
number of alarming trends. For example, the sample showed a reduction of
23 per cent in hedgerow length between 1980 and 1994, and increases in the
built-up area and in coniferous woodland of 4 per cent and 5 per cent
between 1984 and 1990. Ministers were so concerned that they called for
partial resurveys, notably of hedges and ponds, before the next planned
resurvey in 2000.

In Scotland, Scottish Natural Heritage published *The changing face of
Scotland: 1940s to 1970s, Countryside Around Towns Review*, and *Scottish
Farm Income Trends and the Natural Heritage*. The Macauley Institute
published *The Land Cover of Scotland 1988* based on air photographs, and
Scottish Natural Heritage produced *The Islands of Scotland: A living marine
heritage*.

Returning to England, the Countryside Commission made further progress
towards achieving its 'Landscape Character Programme' when it published
*The New Map of England: A Directory of Regional Landscapes: Results of a
Pilot Study in SW England* (CCP445) and *The New Map of England: A
celebration of the South Western Landscape* (CCP444). These divide South-
west England into 38 landscapes which are then described briefly. Both
documents contain similar information, but CCP445 is the layman's version
complete with picture-postcard views, while CCP444 is in the form of a
loose-leaf binder for professional use.

The Countryside Commission also continued to publish its series of
assessments of AONB landscapes mainly carried out by consultants: for
example: CCP424 *The Dorset Downs, Heaths and Coast*; CCP425 *The
Malvern Hills*; CCP442 *East Devon*; and CCP448 *The Isle of Wight
landscape*, the nineteenth in the series. Finally, the Commission issued a draft
policy statement *Views from the past: historic landscape character in the
English countryside*, which discusses how the historic dimension may better
be incorporated into landscape work.

Other landscape-related topics were covered by Clifford, S. and King, A.,
*Local Distinctiveness. Place, particularity and identity*; English Nature, *The
handbook for Phase 1 Habitat Survey — a technique for environmental
audit*; Young, R.H., *Landscape Ecology and Geographical Information*

*Systems*; RSPB, *A shore future: RSPB vision for the Coast*; and Cannell, M. and Pitcairn, C. (eds), *The impact of the hot summers and mild winters in the UK 1988–90*, which examines the potential effects of this period on the flora and fauna.

### Recreation and tourism

*Policy and advocacy* The Countryside Commission in *Delivering country-side information: A good practice guide for promoting enjoyment of the countryside* (CCP447) provided advice for local authorities and organisations on this often-neglected aspect of recreation planning. Another neglected group were also the subject of CCP439, *Informal countryside recreation for disabled people*, a revised edition of a previous booklet. These documents were set in the framework of an established policy as set out in 1992, in CCP371. A similar framework was provided for Scotland, when in October 1994, Scottish Natural Heritage issued *Enjoying the Outdoors: A Programme for Action and Recreation and the Natural Heritage*, following a major consultation exercise in 1993. The overwhelming conclusion of the review was that arrangements for access to the Scottish countryside need improvement urgently, with the most urgent issue being the need to improve local access. Conversely the Scottish Office published a *Guide to measures available to control the recreational use of water*.

In the field of advocacy Curry, N. in *Countryside recreation, access and land use planning* advocates a more market-related approach to recreational provision, while a report for the CPRE by the Centre for the Study of Environmental Change at the University of Lancaster on *Leisure Landscapes* highlights the growth of non-traditional leisure uses (for example, war games) and demonstrates how these clash with prevailing white, middle-class aesthetic values based on the concept of quite enjoyment. The report calls for a major public debate, which hopefully will be created by the 1995 White Paper on the countryside and will be informed by a survey of how 7500 spend their leisure time which was carried out for the Countryside Commission during 1994, and will be repeated in 1996.

*Access* Although not specifically concerned with recreation the Criminal Justice Act and Public Order 1994 could have some implications for access, even though it was mainly directed at hunt saboteurs, New Age Travellers, and the organisers of rock festivals and raves. In particular, the Act introduced the concepts of 'collective trespass' and 'aggravated trespass', both of which will allow the police to remove or arrest those committing either offence, and so in theory traditional access to the countryside should not be affected. Nonetheless, any legislation that would have probably stopped the famous mass trespasses of the 1930s represents a potential threat and thus the way in which the Act is implemented will be closely monitored by PIRPAP.

Meanwhile, an attempt to ensure public access to the countryside for the purpose of open-air recreation, subject to sensible restrictions, was made in the Freedom to Roam (Access to Countryside) Bill 1994 (HC 78(93–94). This Private Members Bill did not, however, proceed past the first reading stage. The Labour Party did pledge themselves to a 'right to roam' which would be enacted in a John Smith Memorial Act, in memory of the former Labour leader who died in May 1994 and who had taken up hill walking after his first heart attack.

In order to help fulfil the pledges made in 1987 in *Policies for enjoying the countryside* (CCP234) the Countryside Commission launched the 'Milestones approach' in *National Targets for Rights of Way* (CCP435) *A guide to the Milestones approach* and (CCP436) *The Milestones approach* which provides a stage-by-stage approach to help local authorities meet the target of 120 000 miles of fully operational rights of way set for the year 2000. Those needing legal advice can turn to *Garner's Rights of Way*, which went into a sixth edition in order to incorporate changes made in the Rights of Way Act 1990. In more detail CCP449 provides advice for farmers and the public in *A guide to procedures for public path orders* and CCP380 provided revised advice on the *Parish Paths Partnership*. Continuing the theme of partnership, the Countryside Commission, NFU and CLA got together to produce CCP450 *Managing Public Access: A guide for farmers and landowners*, which sets out in plain language most of the issues likely to arise. Looking forward, CCP443 *Access payment schemes: discussion paper* sought views over ways in which schemes like Countryside Stewardship which include payments for providing access could be developed. Finally, a new long-distance path, the Midshires Way, was opened, and approval was given for the Hadrian's Wall National Trail.

In Scotland a National Access Forum was established in October 1994 in order to provide a forum for all those involved to discuss promotion and provision of open-air recreation. The Forum will be able to refer to several publications by Scottish Natural Heritage, for example: a *Review of Rights of Way Procedures* by Rowan-Robinson, J.; who with others also produced *Public Access to the Countryside: A Guide to the Law, Practice and Procedure in Scotland*, which is a useful compendium for planners; and *Footpaths and Access in the Scottish Countryside*, written by Peter Scott Planning Services.

*Sport*  Advice on where best to build *Golf courses in the countryside* was provided in CCP438 which expressed a preference for degraded landscapes or areas of intensive agriculture where a new course could make a positive contribution. There should be a general presumption against developments in areas of good landscape. In Wales the Countryside Council for Wales and the

Sports Council in *A Sporting Chance for the Countryside* attempt to show how sport and nature can co-exist by examining good practice in action in nine case studies.

*Tourism*    There were no major developments but a series of progress reports or plans. For example, the latest edition of the English Tourist Board's *English Heritage Monitor 1994* as ever provided all sorts of useful information. A new publication, the *Journal of Sustainable Tourism*, was launched, and the Wales Tourist Board produced a strategic framework for *Tourism 2000* which argued that Wales' greatest asset, the wildness of its landscape, should be safeguarded from developments (for example, windfarms).

## Town and Country Planning

### Plan making

*Improving the local plan process*    A consultation paper with this title was issued by the DOE in April 1994 following a survey which showed that even by the end of 1996, only 91 per cent of authorities would have an adopted district-wide local plan. The DOE was worried that the benefits of the new plan-led system set up by the Planning and Compensation Act 1991 would not be achieved unless plans were put in place quickly. Accordingly, the consultation paper proposed, *inter alia*, removing excessive detail from plans and more effective strategic consultation. However, most responses argued that opportunities for speeding up the process were limited. A second survey published in August 1994 showed that 20 per cent of districts had a plan in place by 31 March 1994, and a map showing the location of these plans was published in *Planning* on 2 September 1994 on p. 8. The process of adoption should be helped by the publication of a *Good Practice Note on Local Plan Preparation* by the Association of District Councils and District Planning Society in July 1994 which gave the process top priority.

*Environmental aspects*    In November 1993 the DOE published *Environmental appraisal of development plans: a good practice guide*, which prescribed clear measures by which planners can develop a basic approach to appraisal which incorporates a number of stages: a check on the environmental content of the plan; examination of the likely impact of policy and proposal options; and modifying policies and proposals to achieve the least harmful impact and the maximum achievement of environmental objectives. The heart of the appraisal approach is to test possible policies and proposals according to their impact on the environment, using a set of criteria to represent the environment at the global and local level. The guide in essence thus gives advice on how to carry out the process of Strategic Environmental

Assessment advocated by the EU at the local level. Further advice to planners about how to translate environmental concern into plan making was provided by English Nature in *Planning for Environmental Sustainability*, Friends of the Earth issued *Planning for the Planet – Sustainable Development Policies for Local and Strategic Plans*, and the Local Government Management Board published *A Framework for Local Sustainability* which was also a response by UK local government to the UK government's first strategy for sustainable development (see p. 33). Previous practice was considered in three publications. First, Healey, P. and Shaw, T., in *Treatment of the Environment by Planners — Evolving Concepts and Policies in Development Plans*, found that economic priorities eventually gained precedence over environmental issues as plans were developed. Second, the CPRE in a survey of 70 plans accused planners of key *Environmental Policy Omissions in Development Plans*. In contrast, the third study by Therivel, R. (ed.), *Environmental Appraisal of Development Plans in Practice*, found that many more local authorities were carrying out appraisals six months after the publication of the DOE advice, and that good practice was generally being followed, albeit mainly based on primarily descriptive unweighted matrices.

*Other specific advice* In December 1993 the DOE issued a consultants' report on *Alternative Development Patterns: New Settlements*, which argued that there was a strong case for new settlements to be built and that they should be identified by plans rather than by developers. The DOE used the report to reiterate the advice in PPG3 that new settlements were acceptable where the continuing expansion of towns and villages would be a less satisfactory method of providing land for new housing. In 1994 the DOE continued the theme of encouraging growth back into towns when it published *Vital and viable town centres: meeting the challenge*. Economic issues were the concern of specific advice from the RTPI in *Local authority-wide economic development plans*, a practice guidance note for its members. On a wider canvas the Countryside Commission published a *Countryside Planning File* (CCP452), which summarised all the information published by the Commission on planning-related issues, and internationally the Food and Agriculture Organisation published *Guidelines for land-use planning*. Advice on how to deal with data was provided by the County Planning Officers Society in *Information and Computing for County Planning*, while the specific case of population data was covered by Dale, A. and Marsh, C. (eds) in *The 1991 Census Users Guide*.

*The inquiry stage* In November 1993 a consultants' report into *The Efficiency and Effectiveness of Local Plan Inquiries* was issued by the DOE. It found that the time taken between depositing the plan and the receipt of the inspector's report had doubled over the last ten years, and recommended fundamental changes if this time were to be reduced. However, the ADC in their response commented that the piecemeal changes proposed were likely to bring about only minor improvements, and the DOE seemed equally lukewarm over the proposals. Meanwhile, March 1994 saw the last structure plan (Devon) to be approved by the DOE, while earlier in January, Leicestershire became the first authority to self-certify its structure plan with 4000 fewer houses than recommended by the Examination-in-Public Panel. Finally, the Peak National Park Authority voted in December 1993 to reject DOE advice to modify its new Structure Plan.

*Regional plans*   More Regional Planning Guidance (RPG) was published by the DOE during the year, for example: RPG *Northern Region*; RPG8 *East Midlands*; RPG9 *South East*; and RPG10 *South West*. Draft guidance for the West Midlands was issued in September 1994. In Northern Ireland, *A Planning Strategy for rural Northern Ireland* endorsed the policy of favouring single houses in the countryside which makes Northern Ireland a pariah or pioneer depending on your viewpoint.

*Development control*

*Regulatory change*   In December 1993 the DOE issued three consultation papers, under the general title 'Streamlining Planning', aimed at reducing the number of decisions on minor applications with respect to commercial premises, outdoor advertisements, and where planning controls are duplicated by other controls. The proposals were put into operation in March by SIs 678 and 724/94. In contrast, in March 1994 the DOE announced that it would be removing crown exemption from planning permission as soon as a legislative opportunity arose. In May, however, a Department of Trade and Industry white paper on *Competitiveness — Helping Businesses to Win* noted that there was still room for improvement in planning procedures, notably by introducing so-called fast-track procedures.

*Fees*   Fees for planning applications rose by 15 per cent in January 1994 (SI 3170/93) with a further 15 per cent predicted for 1995, in an attempt to make applicants meet the full costs of the service they use which, at present, ranges from 23 per cent to 143 per cent according to a survey of 25 authorities carried out for the DOE. In June 1994 a DOE consultation paper proposed that councils could set their own fees as from 1996, and also the idea of so-called 'fast-track' applications in return for a higher fee, which had already been floated in the May White Paper on 'Competitiveness'. The proposals were greeted with universal horror from across the spectrum of interested groups, including the RTPI, the CPRE, and the NFU, with fears of high fees being charged under monopoly control, and that the proposals would undermine the integrity of the planning permission.

*Role of the Countryside Commission*    In November 1993 the Countryside Commission in a statement (CCP415) announced that it would no longer comment or become involved with development control matters unless it set a national precedent where national government guidance is lacking; would have a major impact on a national Commission initiative; or have a fundamental impact on National Park, AONB, Heritage Coast or equivalent area. The Commission thus set itself a selective and strategic role.

*The appeals and inquiry system*    The Planning Inspectorate Executive Agency published its first *Annual Report and Accounts for the year ended 31 March 1993*. This showed that the great majority of the targets set in its first business plan had been achieved, albeit helped by a falling case-load. The *Second Annual Report and Accounts* published in September 1994 revealed continued improvements with the best-ever performance, at least in time taken to reach decisions, ever achieved. The Agency also published a *Corporate Plan for 1994–97* and a *Business Plan for 1994/5* which set out where it expected workloads to rise and fall. In another forecast *The Enforcement Trends Study* predicted a 10 per cent growth in such work over five years. In December 1993 the DOE published the results of a consultants' *Study of Appellant's Experience of the Written Representations Appeals System*, which showed that they were very satisfied with the procedure. Nonetheless, a number of recommendations were made for change, most of which were implemented by the Inspectorate. In June 1994 a similar report on the 'hearings procedure' was issued which made six recommendations. The Inspectorate agreed to take action on three of these, but rejected the main recommendation that the 'hearing method' should be more widely used. Advice on how to use the system was provided by Banks, S. and Casley-Hayford, M. in *Practical Planning Appeals and Inquiries*. Turning to the special case of roads inquiries, the Department of Transport announced in August 1994 the implementation of 'round-table' conferences at which contentious road schemes would be discussed in the hope of reaching consensus. The conferences will not replace statutory public inquiries, nor reduce the rights of objectors, but it was hoped they would shorten decision times, reduce the period of uncertainty, and help speed up delivery of the road programme. The conference procedure had been tested on the Hereford Bypass and the A30 in Cornwall. A consultation paper outlining the procedure and other methods for reducing the time taken to build roads from 13 years to eight had been issued in March.

*Planning Policy Guidance Notes (PPGs) and circulars*

*PPG13 Transport*    The continuing retreat from car-based planning policies was marked by this PPG, published in March 1994, which strengthened the policies contained in the draft. The central messages though remained the same, namely: development that attracts journeys must be located where it is accessible by a choice of transport modes; dispersed facilities and edge-of-town locations should be disapproved of; and housing schemes must be accessible to public transport. In the same month the roads programme was substantially scaled down in an effort to reduce fuel consumption and emissions by at least 15 per cent over the next 20 years. However, by the end of the summer a conference on PPG13 was able to conclude that although it was a visionary document, many remained doubtful about its real ability to deliver substantial changes. Nonetheless, the year saw a number of out-of-town developments turned down after appeal by the DOE, and several retailers, including Tesco, announced a renewed focus on urban outlets, in a response to not only PPG13 but PPG6 which advocated fewer out-of-town stores in 1993. Finally, research in the form of The Merry Hill Impact Study demonstrated that this out-of-centre shopping development had had a severe impact on nearby Dudley town centre.

*PPG23 Planning and Pollution Control*   The main point made by this July 1994 PPG is that the planning system should not duplicate controls which are the statutory responsibility of other bodies, and should assume that the pollution control regimes will be properly applied and enforced by them. There may be cases, however, where planning's wider remit for protecting the environment may take precedence, but only on land-use grounds. The PPG was thus criticised for failing to clarify a murky area.

*Draft revision of PPG2 Green Belts*   In February 1994 the DOE issued a draft revision of the 1988 PPG2 on Green Belts. The main proposals were: to set out positive objectives for land use; to encourage proper consideration of the long-term direction of development; to modify the categories of new building considered appropriate; to make realistic provision for the future of existing employment sites; and to permit existing buildings to be re-used. The draft thus reaffirmed existing advice, following research by Martin Elson for the DOE which showed that existing policies were working well. Not surprisingly, Martin Elson argued that the draft was an opportunity seized, and the RTPI, the TCPA, and Countryside Commission broadly welcomed the thrust of the draft. The TCPA and CPRE did, however, express reservations about the supposed link between restrictive policies and urban renewal, expressing concern that so-called 'town-cramming' could be the result. In October 1994 the Secretary of State for the Environment announced that the revised PPG would further strengthen Green Belt Policy.

*Summary of PPGs and consultation papers*   A useful one-page guide and list of consultation papers and PPGs as at April 1994 was set out on page 23 of the 22 April 1994 edition of *Planning*.

*Circulars*   Although circulars have largely been superseded by PPGs a few are still issued, albeit on specific issues. For example, Circular 1/94 *Gypsy Sites and Planning* reminds local planning authorities that development plans should have clear policies on gypsy site provision, even though the Criminal Justice and Public Order 1994 removed the duty of local authorities to provide sites, and for central government to provide grant aid for such sites. Elsewhere, the Department of Transport issued a Circular on *Motorway Service Areas* (1/94) in the context of deregulation in 1992, which has seen developers gain planning permission for five sites, with 25 sites being sought approval.

## Other planning issues

*Environmental assessment*   In April 1994 private toll roads were made the mandatory subject of an environmental assessment, while wind generators, motorway service areas, and coastal protection works were made the potential subject of an environmental assessment if local authorities so request (SI 677/94 and Circular 7/94). However, golf courses, water treatment plants, trout farms, and major service areas on non-motorway roads which had been promised to be included were dropped, at the last minute, in March, much to the horror of the CPRE. This put the DOE on a collision course with the European Commission, which published draft amendments to the environmental assessment directive in April 1994. These included proposals to widen the scope of Annex 2 projects to cover various types of tourism-related development, among them golf courses and skiing developments, and where projects are likely to have a significant effect on the Special Protection Areas to be set up under the Habitats Directive. The House of Commons Select Committee on European legislation in a report on the draft amendments noted the disagreement between the Commission and the government over certain aspects of environmental assessment and promised to consider the matter again (HC 48–xii(93–94)). Later in the year the DOE reported that research conducted for them had found that environmental assessments submitted by developers in support of planning applications were inadequate. Accordingly, the DOE issued a draft *Guide on Preparing Environmental Statements for Planning Projects* which was intended to complement earlier advice published in *Environmental Assessment — A Guide to the Procedures*. The DOE also issued *A Guide to the Eco-Management and Audit Scheme for UK Local Government* to help local authorities follow environmental practices in their own work which will be an example to others. Advice is also available in Glasson, J., Therivel, A. and Chadwick, A., *An Introduction to Environmental Assessment*, which is an excellent book for anyone wishing to know what Environmental Impact Assessment (EIA) is and how to practise it. It is, however, less good as a critique of the world in which EIA has to operate, and especially the realities of power. For those interested in the progress of a technique that is growing increasingly influential in planning this book should be required reading, and even for those whose only interest in planning is to decry it as an agent of capitalism, it will be a reminder that planning does have robust and meaningful tools. There have been too few books recently about how to practise planning, and it is to be hoped that this will lead to a revival. In conclusion, this is a readable, well-illustrated, and affordable text that I will find indispensable for teaching and writing, and it is to be hoped in due course that the offspring of EIA, Strategic Environmental Assessment, which is briefly included in the closing chapter, will form the subject of a second volume before the decade ends. Erickson, P.A. provides *A Practical Guide to Environmental Assessment* from an American perspective.

*Design*   The DOE issued a discussion document on *Quality in Town and Country* which called into question the cost of past planning policies based on disparate zoning of different everyday activities, and outlined ways of reducing the pressure for urban sprawl. It accepted that development in the countryside would still be needed, but that this could still be accommodated in ways which would sustain distinct rural communities. More detailed ideas on *Design in the countryside* were provided by the Countryside Commission in CCP418. This argued that diversity and distinctiveness in design are the ways in which one place is distinguished from another. The theme of regional diversity thus offers the opportunity to achieve local distinctiveness without recourse to pastiche or giving way to relentless standardisation. One method by which design standards can be implemented are through development control, and so in *Countryside Benefits from Planning Obligations* (CCP440) the Commission outlines how planning obligations can enhance the beauty of the countryside. The particular issue of *Environmental factors in road planning and design* was considered by the National Audit Office in HC 389(93–94) which examined 12 Environmental Statements that had been prepared for new road schemes. It was concluded that these complied with the European Directive, were well presented, and described the development of their settings well. However, improvements could be made in the assessment of certain key aspects and the effectiveness of mitigation measures.

*Protecting the heritage*   In March 1994 the DOE announced that it planned to introduce an amendment to the GDO which would allow local authorities to withdraw permitted development rights in self-selected conservation areas, giving them greater control over works affecting external appearance — albeit only for houses (for example, doors and windows). The announcement was a partial reaction to a consultation paper on tightening conservation area controls issued in July 1993 which had been generally criticised for not going far enough during the winter of 1993–4. In particular, an RTPI commissioned study, *The Character of Conservation Areas*, concluded that though they had protected the heritage from major change too many areas lacked proper management, funding, and a clear sense of purpose. Major legislative change was not, however, recommended, only the more effective use of existing powers. More general draft guidance issued the previous year as draft PPG15 *Historic Buildings and Conservation Areas* was criticised by the Countryside Commission for paying insufficient attention to the countryside, especially since in some areas more than half of all villages have been designated as conservation areas. The Commission also called for a new form of designation to protect rural features. The National Heritage Committee produced a report on *Our Heritage: Preserving it, prospering from it* in HC 139(93–94) which focused on the work of English Heritage. The report found a serious lack of coherence about policy for the preservation of our heritage and its very important links with the tourist industry. For example, the Department of National Heritage grant aids 34 quangos, and so the Committee recommended that the Department of National Heritage should publish a clear policy on marketing, funding and, above all, on priorities. In their response (HC 549(93–94) the government argued that its priorities and funding policies had been clearly stated, but that it would continue to work to achieve better coordination and integration of policies. In more detail, the government delayed a decision on a recommendation to introduce a statutory requirement on planning authorities to consult English Heritage on planning applications affecting registered sites of historic gardens and parks, along the lines of that already introduced in Scotland. In response to a recommendation concerning a lower rate of VAT for the repair and routine maintenance of listed buildings the government said that this was a matter for the EU.

The government also responded to a Welsh Affairs Committee report on *The preservation of historic buildings and ancient monuments* in Cm 2416 by reiterating the message that in caring for the built heritage there is a need to balance competing priorities. The responses concerned the details of these priorities but did not involve any major change in policy.

The expenditure plans for National Heritage were set out in Cm 2511, as well as its second annual report which covered its three broad areas of activity: preserving the heritage of the past; creating the culture of today and adding to the heritage for future generations; and broadening opportunities for people to enjoy the benefits of their heritage and culture. These areas of activity contain no specific reference to rural areas, but rural recreation provision and the funding of environmental projects in the countryside fall within the remit of the Department, and its commitment to oversee the dispensation of funds from the National Lottery which had its first draw on 19 November 1994. As part of this work the Department continued the task of listing listed buildings formerly undertaken by the DOE. In mid-1993 the list contained 442 042 entries.

Meanwhile, the more narrowly conceived English Heritage in its *Annual Report and Accounts 1992/93* recorded an increase of 14 per cent in grant aid for listed buildings to a record £33 million. Elsewhere, Historic Scotland presented its *Annual report and accounts for 1992–93 and 1993–94* in HC 910(92–93) and HC 528(93–94). The second edition of Mynors, C., *Listed Buildings and Conservation Areas* was published. Revised restrictions intended to limit farmers' freedom to plough over ancient monuments were proposed in November 1993 and implemented in June 1994 in SI 1381/94. Ancient monuments were also the subject of *Annual Reports for 1993* from the Ancient Monuments Board for Scotland in HC 358(9–94) and a *Report for 1992–93* by the Royal Commission on the Ancient and Historical Monuments of Wales. More generally, the Council for British Archaeology in *The Past in Tomorrow's Landscape* called for new and urgent measures to protect the heritage from further destruction, and criticised the lack of a coherent policy for the care and interpretation of Britain's landscape. Finally *Desiderata: The International Journal of Heritage Studies* was launched in late 1993 by the University of Plymouth's School of Design based in Exeter under the editorship of Peter Howard.

*Minerals and energy*    The Coal Industry Act 1994 (Chapter 21) established the legal framework for a privatised industry. It also provided for a Coal Authority to be set up to issue licences to private sector mining companies, and to take over responsibility for dealing with the consequences of historic coal mining, such as subsidence or water pollution. The NFU expressed concern over several aspects of the Act, notably the possibility that thousands of hectares of farmland could be flooded or contaminated by abandoned mines. MPG3 *Coal mining and colliery spoil disposal* replaced the 1988 MPG3 and Interim Planning Guidance issued in March 1993. It removed a 'national interest' test which had been contained in a December 1993 consultation draft, brought opencast mining more firmly under mining planning authority control, but cautioned the planning system against setting production limits, since these were matters for operators to determine. Opencast coal production has risen from 8 million to 19 million tonnes between 1945 and 1992, and so the Countryside Commission produced *Opencast Coal Mining: advice on landscape and countryside issues* (CCP434) in an attempt to provide a more strategic overview, further improvements in restoration, and long-term management of restored landscapes. The growth of opencast is also portrayed in County Planning Officers Society, *Opencast Coalmining Statistics 1992/93*. MPG6 *Guidelines for aggregates provision in England* set a target of reducing the proportion of building aggregates taken from land-won sources to 68 per cent by 2006 compared to the 1992 figure of 83 per cent. The MPG also called for more recycling, a reduction in land banks from ten to seven years, and reducing the government's planning horizon from 2011 to 2006. The MPG, which followed a January 1993 consultation paper, was criticised by operators and conservationists for opposite reasons, but the County Planning Officers Society welcomed the advice. MPG 12 *Treatment of disused mine openings and availability of information on mined ground* provided advice on how to deal with the major hazard posed by the estimated 250 000 abandoned mine openings. In March 1994 the DOE issued a consultation paper *The Reform of Old Mineral Permissions 1948–81* which proposed ways of regulating permissions obtained in this period with those obtained before and after. A similar paper was issued by the Scottish Office in July 1994. Information about the related issues of landfill, derelict land reclamation, and waste disposal was provided in five publications: Department of Environment, *Assessing the effectiveness of derelict land grant in reclaiming land for development*; Warren Spring Laboratory *et al.*, *Externalities for Landfill and Incineration*; the House of Lords Select Committee on the European Communities response to the European Commission Green Paper on *Remedying Environment Damage*, HL 10(93–94); National Audit Office, *Land reclamation in Wales*, HC 461(93–94); and County Planning Officers Society, *Planning for quality in minerals and waste development control*.

The Department of Trade and Industry discussed *Energy* and the Energy Technology Support Unit made *An assessment of renewable energy for the UK*. These both concluded that geothermal and tidal power were limited in their potential, and that coppice, waste, wind, solar and fuel cells, landfill gas, and small-scale hydro presented the best prospects, but that planning procedures sensitive to the needs of local communities would be required if these renewable technologies were to fulfil their potential to improve the global environment. More specifically, *Wind energy* was discussed by the Welsh Affairs Committee (HC 336(93–94)) and the Countryside Council for Wales published research from a *Wind Turbine Power Station Monitoring Study*, which revealed that the visual impact was considerably more than anticipated, with turbines being clearly visible up to ten miles away. Nonetheless, Friends of the Earth argued that public opinion polls showed strong majority support for wind power and so they published advice on *Planning for Wind Power*. Finally, the Countryside Commission issued a position statement on *Overhead Electricity Lines: Reducing the Impact* which called for the introduction of targeted undergrounding.

*Reviews of organisational performance*  The DOE's *Departmental Report and its Expenditure Plans for 1994–94 to 1996–97* were published in Cm 2507 which set out in useful detail its work under ten chapter headings including: Environmental Protection; Countryside and Wildlife; and Planning. It also described the achievements and developments made by the DOE in the previous year, which makes it a very useful source of information on policy developments and expenditure patterns.

The work of the DOE as revealed by Cm 2507 was examined by the House of Commons Environment Committee in HC 414(93–94), which concentrated on housing policy. However, it also pointed out that the report contained no reference to the aims or achievements of the Joint Nature Conservation Committee. Accordingly, the Committee recommended that summaries of the aims and achievements of all divisions of the DOE should be included in the Annual Report as a matter of course.

Also at the national level the DOE published the 16th *Digest of environmental protection and water statistics*. In August 1994 the DOE issued the third in the series of six monthly *Planning Performance Checklists*, as part of the Citizen's Charter initiative. This ranked authorities according to the percentage of applications decided within 8 weeks, although the government points out that speed and quality of decision making should be taken together. The quality of decision making is, of course, impossible to measure, but good indications are given by the discussion of planning issues raised by those who complain to the Local Government Ombudsman and outlined in his *Annual Report 1993–94*. The checklist was augmented in September by publication of the *Statistics of Planning Applications April to June 1994*. Data on land use changes in 1989 were published by the DOE in the ninth *Land Use Change Statistical Bulletin*. This revealed a net increase of 5600 ha in urban land over rural land, but also that 47 per cent of land developed for urban uses was previously developed for urban uses which was part of a growing trend to recycle urban land rather than developing greenfield sites. However, the South East Regional Research Laboratory in *Analyses of Land Use Change Statistics* warned that, although useful, the statistics also had severe limitations. The DOE also published research into *The cost of determining planning applications and the development control service* and another in the series of *Local Government Financial Statistics 1990/1–1993/4* which provides detailed data on the costs of providing planning services. November 1993 saw the publication of Audrey Lees' *Enquiry into the planning system in North Cornwall District* which had been the subject of allegations about malpractice. The Land Authority for Wales, the residual body from the 1970s experiment into public land assembly, published its *Accounts for 1993–94*, and also in Wales the Welsh Office issued a booklet on *Planning Charter Standards* under the Citizen's Charter initiative. Finally, the Town and Country Planning Association in its *Annual Report 1993–94* gave precedence to the publication of its report *Planning for a Sustainable Environment* and its widespread impact.

## Town and country planning in Scotland

*Review of the system*   A consultation paper on the system was issued in July 1994 by the Scottish Office. It argued that the general structure of the system was sound, and that it had broadly succeeded in facilitating development while protecting the environment. Accordingly respondents were directed towards a number of specific questions relating to procedural rather than strategic matters.

*National planning policy guidelines*   A number of new NPPGs were issued during the year. The most important was NPPG1 *The Planning System*, which gave clear objectives for the planning system and the primacy of development plans. Three objectives were set for the system: to set the land-use framework for promoting economic development; to encourage economic, social and environmental regeneration; and to maintain and enhance the quality of the natural heritage and built environment. It also underwrote structure plans as the strategic level of planning, and reminded local authorities that local plans must not only be produced by the end of 1995 but after that they should be kept regularly up to date. Advice on other issues was issued in: NPPG2, *Business and Industry*; NPPG4, *Land for Mineral Working*; NPPG5, *Archaeology and Planning*; and NPPG6, *Renewable Energy*.

*Planning Advice Notes*   A number of new PANs were issued during the year: PAN41 on *Development Plan Departures*; PAN42 on *Archaeology*; PAN43 on *Golf courses and associated developments*; PAN44 on *Fitting New Housing into the Landscape* which was complemented by a report by the Scottish Office Building Directorate on *Timber Frame Houses in the Scottish Countryside*; and PAN45 on *Renewable Energy Technologies*.

*Circulars and Statutory Instruments*   A major revision of Circular 17/87 was made be Circular 27/93 in an impressive 134-page document which provided a *Memorandum of Guidance on Listed Buildings and Conservation Areas*. Circulars 3, 13 and 16/94 provided specific information on departures from the Development Plan as set out in NPPG1 and PAN41 (*ante*) and on a Direction which took effect on 7 March 1994. Fees for planning applications were raised by SI 3211/93 and described in Circular 1/94. Circular 25/94 raised the threshold above which applications to develop prime quality agricultural land have to be referred to the Scottish Office Agriculture Department from 2 to 10 ha. Circular 26/94 drew attention to SI 2012/94 which added private toll roads, wind generators, motorway service areas, and coast protection works to the list of projects that either must have (toll roads) or might be subject to an environmental assessment. Finally, A *review of the Use Classes (Scotland) Order* was carried out by Brand, J. and James Barr and Sons for the Scottish Office.

*Other information on planning in Scotland*   This is provided by the Scottish Office, *Literature Review of Rural Research Issues*, and Bryden, J. *et al.* in

*Pounds, Policies and Prospects*, which details the ways in which EU spending, policies and legislation affects rural Scotland.

*Textbooks, other reviews and Guides on Town and Country Planning*

The main event of the year was the relaunching of Cullingworth's *Town and Country Planning in Britain*, which until Rydin's excellent new book was published in 1993 (see Volume Four, p. 134) had been the standard, if dull, text for British town planning. This much-modified eleventh edition, which sees the introduction of Vincent Nadin as a co-author, also witnesses two other new introductions: first, the insertion of a more discursive approach, and second, a widening of the subject matter to include a new chapter on 'Heritage Planning'. Conversely, the chapter on 'Regional Planning' has been dropped. Other chapters have been rejigged: for example, 'Planning for Traffic' becomes 'Transport Planning', in a reflection of changed attitudes to the environment and its planned management in the last decade. In general, the changes work and I found myself ordering several new slides for use in my lecture courses, whereas previously Cullingworth had been as telegenic as a 1960s White Paper. The bibliography is also now vastly improved and the official publication section is now excellent, and here the contribution of 'PIRPAP is fully acknowledged in the reference list. However, the book remains weakish in the area of development control, and on the impacts of the system. In addition, it lacks a discussion of the place of planning in society, or any all-embracing theory of how and why planning operates at all. To some extent, Rydin's book went too far in this direction and so the two books together make a splendid combination for students wishing to get to grips with the very wide field of planning as it is practised today. As a descriptive stand-alone account of the skeleton of planning in Britain this edition re-establishes Cullingworth as the market leader, and it would be a brave author or publisher who would try to do better. Other texts include: Adams, D., *Urban Planning and the Development process*; Birtles, W. and Stein, R., *Planning and Environmental Law*; Brand, C., *Planning Law*; Casley-Hayford, M. and Banks, S., *Practical Planning: Permission and the Application*; Emmett, S., *Practical Planning: Enforcement*; and Tromans, S. and Grant, M., *Encyclopaedia of Environmental Law*. A more general *Guide to Effective Participation* by Wilcox, D. shows how people can participate in all forms of community participation, not just planning. Finally, *The Impact of the European Community on Land Use Planning in the UK* is considered by Davies, L. and Gosling, J. in a report for the RTPI. This identifies three predominant forces: first, the creation of the single market which will bring changing economic pressures; second, increasing attention being given to sustainable development and the quality of life; and third, more subsidiarity so that decision making becomes more devolved throughout Europe but within a European rather than national framework.

*Theory and values*    The search for a theory of planning that will suit both theoreticians and pragmatic practitioners may seem a fruitless one, but one that your editor is struggling with, much to his cost. Several new contributions were made during the year, most notably, Forester, J., *Critical Theory, Public Policy and Planning Practice: Towards a Critical Pragmatism* which received a very positive review in the *Journal of the American Association of Planners*, 13(3), 1994, 229–30. Other contributions were made by Fischer, F. and Forester, J. (eds), *The Argumentative Turn in Policy Analysis and Planning* and by Sager, T. in *Communicative Planning Theory*. All three of these books focus on Habermas' validity claims in communication: truth; rightness; truthfulness; and comprehensibility as they relate to planning practice. In a different vein Thomas, H. (ed.), *Values and Planning*, examines aesthetic judgement, environmental ethics, and the significance of key concepts like heritage, justice, professional ethics and the public interest.

## Social and economic issues

*Community and services*

In February 1994 the DOE launched 'Rural Challenge' a new competition designated to stimulate innovative approaches to social and economic development in less prosperous rural areas. There will be six prizes each year of £1 million to be spent over a three-year period. Bids will be made by partnerships of any group or organisation from the public, private or voluntary sector, including the Rural Development Commission's Rural Development Area. In order to fund the scheme the Commission's budget was raised from £38 million to £43 million but it was expected that successful bids would contain a high proportion of private sector investment. An earlier scheme, 'Rural Action for the Environment', which was launched in 1992, had funded nearly 300 community projects at a cost of over £1 million and had grown to cover 32 countries by February 1994. Rural Action was also the title of a book edited by Henderson, P. and Francis, D., *Rural Action — a collection of community work case studies*. This book did not, however, consider the scheme outlined above, but ten case studies across Europe. These illustrated essential principles which transcend local circumstances and thus inform practice regardless of location.

Turning to services, the Rural Development Commission issued *Rural Services: Challenges and Opportunities* which called for action on three fronts. First, the planning system when considering new development should assess the beneficial effect on services that might acrue. Second, public policy makers should give more weight to rural circumstances when planning service provision. Third, new ways of meeting local needs should be found. The Post Office was one service that was put under threat during the year following proposals to privatise it in Cm 2614, *The future of postal services: a consultative document*, but following a backbench revolt by about a dozen Tory MPs the government withdrew the proposal in October 1994.

*Countryside concepts and lifestyles*

The first in-depth study of *Lifestyles in rural England* was carried out by Cloke, P. and Milbourne, P. for the Rural Development Commission. From a survey of 3000 households in 1991 they found a diversity of employment and income groups. Some areas were characterised by middle-aged people on low incomes happy to trade-off poor services for perceived quality of life, while in other areas, notably near the big cities, affluent workers were the norm. Cloke, P. is also one of five authors behind *Writing the Rural*, which attempts to explore the myriad concepts of just what is rural, because the lack of any commonly accepted definition is proving a real barrier to any attempt to understand (let alone plan) rural areas. Each of the five essays attempts to draw out the 'rural' by drawing on different traditions in social and cultural theory. Another attempt by Marsden, T. and Murdoch, J., *Reconstituting Rurality*, examines the way in which the 'rural' and the concept of rurality is being reconstructed within urban regions, and argues that the 'rural' is not a fixed category but the outcome of political, economic ad sociocultural pressures. From a series of case studies it concludes that certain social groups are becoming increasingly influential in creating a rural locale in the likeness of class groups. Other books that examine these issues are Derounian, J., *Another country: Real life beyond rose cottage* and Bunce, M., *The Countryside Ideal: Anglo-American Images of Landscape*. Finally *Ecumene*, a journal of Environment, Culture and Meaning, was launched by Edward Arnold in 1994 with a £33 subscription.

*Housing*

A general *Inquiry into Planning and Housing* was made by the Joseph Rowntree Foundation. This concluded that the role of the planning system in solving housing problems had been exaggerated and that enough land has been allocated for house-building. Accordingly, it rejected the idea of special areas where planning permission would be restricted to social housing. Further research for the Foundation by Jackson, A. *et al.*, *The Supply of Land for Housing*, examined the new primacy of development plans in making development control decisions. The research did not conclude that this would lead to more land being released since ultimately this is a political rather than a scientific decision, and that the planning system by itself could not generate large amounts of affordable housing. *Planning for affordable housing* was studied by Barlow, J. *et al.* for the DOE. This found that although over 10 000 such houses had been given permission, only around 2000 have been built. Accordingly, the study made nine recommendations which included the view that policy should be led by the assessment of local need, rather than as a response to incremental demand. Affordable housing is often funded by the Housing Corporation and in Cm 2363 the government responded to an earlier report on its work by the House of Commons Environment Committee. However, the recommendations dealt mainly with management structures and finance and contained little of direct relevance to rural planning.

The Rural Development Commission carried out *An Evaluation of the Housing Corporation's Rural Programme* under which 5616 houses were funded between 1990 and 1993. It concluded that sites are obtained adventitiously rather than by a strategic assessment of need and recommends that development control should be more flexible. Nonetheless, the RDC's *Statistical Update 1992–93: Homelessness in Rural England* reported that 16 000 rural households were accepted as homeless in 1992–3. In a response to the 'immense scale of the rural housing problem' Rural Voice in *Meeting Rural Housing Needs* called for social housing targets to be incorporated into all levels of plan making while acknowledging that money rather than the planning system was the main problem. In Scotland a *Review of Support for Crofter Housing* by Scottish Homes recognised the vital role of various support measures but called for them to be incorporated into a one-step system.

*Transport*

The government's *Expenditure Plans for Transport 1994–95 to 1996–97* were set out in Cm 2506 which also provided a useful overview of transport policy and contained a new chapter on 'Transport and the Environment' with a section on the environmental impacts of transport. It also included an outline of the responsibilities of the new Highways Agency which began work on 1 April 1994 on delivering the national motorway and truck road programme. The plans also revealed that expenditure on new roads construction would fall from £1639 million in 1993–4 to £1190 million in 1996–7.

The previous year's report was examined by the House of Commons Transport Committee in HC 722(92–93), which included a recommendation that the next report should explain in greater detail, how the government intended to reduce $CO_2$ emissions in line with the Rio undertakings, to which the government responded in HC 363(93–94) by noting that the new chapter on 'Transport and the Environment' referred to above had filled the gap.

The planned expansion of the road network was substantially cut back by the Department of Transport in *Trunk Roads in England 1994* with 49 projects being scrapped and work on 69 schemes being suspended. Meanwhile, the European Commission outlined its plan for a trans-European network for 2002 which included the upgrading of 12 UK routes with the particular aim of linking peripheral regions. The Association of County Councils in *Towards a Sustainable Transport Policy* advocated locating homes, jobs and services closer together, improving the quality of life in towns in order to slow down the exodus to the countryside, and developing balanced communities along transport corridors with good public transport services. In a similar contribution the CPRE in *Local Moves* also called for a more effective use of public funding by subjecting transport policy to an environmental assessment which would reveal the real cost of private road transport and thus lead to a reallocation of funding to public transport. According to Transnet, a research group, this reassessment should include *New Futures for Rural Rail* in which rural stations could be turned into 'local community enterprise stations'.

The impact of traffic was considered by the Institution of Highways and Transportation in *Traffic Impact Assessment Guidelines*, while the National Trust threatened to fight to the end to prevent a new road impacting on its Golden Cap estate in Dorset.

*Employment*

The Rural Development Commission (RDC) issued a *Rural Development Strategy for the 1990s* which contained a new map of Rural Development Areas which covered more than a third of England and 2 750 000 people, and came into operation in April 1994. Five new Areas were created, all of them south of the Severn–Wash line, but two were discontinued, while seven were expanded and eight were reduced in size. No changes were made in the form of assistance, but support for voluntary activity and services were seen as two key strands in the strategy, as evidenced by the £6 million 'Rural Challenge' initiative described above. The RDC also commissioned a report on *Rural Development and Statutory Planning* which from a case study of seven areas concluded that local planners were blocking business start-ups in contravention of the advice given by PPG7, and that structure plans were more likely to advocate the policies of PPG7 in favour of development than local plans which tended to favour landscape protection. The need for planning authorities to provide more sites for small-scale employment schemes in their local plans was also taken up in the RDC's *Annual Report for 1993–94*.

In Scotland, the Scottish Development Agency was finally dissolved by SI 1976/94 while its successor Scottish Enterprise published its third accounts in HC 589(93–94). Highlands and Islands Enterprise also published their *Third Report 1993–94* which showed employment growth in their area to be above the Scottish average in its most successful year to date. In Wales the Development Board for Rural Wales' *Accounts 1993–94* (HC 642(93–94)) recorded a very successful year with 1840 jobs assisted compared to the target of 1500. However, the economy of the area remains fundamentally weak with GDP only 76 per cent of the EU average.

European aid under Objective 5b was extended from 1.5 million to 2.8 million people in January 1994 in a major review of the areas eligible for aid. Specific proposals totalling some £665 million were submitted to the EU in May for a variety of measures designed to improve rural infrastructure and economic diversification between 1994 and 1999. Seven rural development programmes from the submission were approved by the EU in October totalling £300 million. The European Foundation for the Improvement of Living and Working Conditions considered the potential revolution in our lives to be wrought by *Telelifestyles and the flexicity* and in *Telecottages* Acre examined *The UK Experience*.

## The reform of local government

*England*

The whole of this review could have been filled with the U-turns and general confusion surrounding the process in England. This began with a DOE statement in November 1993 re-emphasising the benefits of unitary authorities and asking the Local Government Commission to give more weight to proposals emerging as a result of local consensus. After this, the year saw a good deal of confusion as draft proposals made by the Commission were often altered or even stood on their head by them by the time they were submitted to the DOE, only to be sometimes altered again by the DOE when they returned them to local people for their comments. In some cases the DOE asked the Commission to think again, as, for example, in Derbyshire. In addition, there were legal challenges to the process which added to the general uncertainty, and the timetable itself was altered, although it was still hoped to complete the process by the end of 1994. As the process progressed, more and more two-tier solutions broadly based on the status quo began to emerge. By October 1994 some sort of recommendation had been made for each county, which if carried through would lead to a very confused pattern of local authorities by the end of the decade. Based on a map produced by *Planning* on 16 September 1994 there would be 15 counties with two-tier systems broadly similar to the present, but with the loss of big cities (for example, Plymouth). The rest of the country would have a unitary system, but not at all one based on existing districts. The future looks very confused and, if anything, this reform is even worse than the ideologically driven 1974 reform.

Commentaries on the process were provided by: evidence submitted and examined by the Environment Committee in *Review of Local Government in England* (HC 922(93–94)); Audit Commission for Local Authorities, *Time for change? A consultation paper on work to support the implementation of local government reorganisation*; and Institute of Local government Studies, *Local Government Review: Key Issues and Choices*.

## Scotland

In Scotland the Local Government Scotland Act 1994 broadly implemented the proposals set out in Volume Four of PIRPAP on p. 101, namely a unitary system of 28 authorities to take power in 1996. The main planning issues raised during the Bill's passage were the arrangements for multi-district structure plans in a totally unitary situation.

## Wales

In Wales the Local Government (Wales) Act 1994 broadly implemented the proposals set out in Volume Four of PIRPAP on p. 101, namely a unitary system of 21 authorities to take power in 1996. The Act set up powers to create joint planning boards for the national parks even where they fall wholly within one council (for example, Pembrokeshire) and also provided for unitary development plans.

## Author's note

This review has been produced from press releases mainly from the MAFF, DOE, and NFU but also from a number of other organisations. It has also been sourced from *Planning, Planning Week*, and *Farmers Weekly* as well as a number of other publications. Government publications have been taken from *The Daily List of HMSO Publications*. Publications are referred to by the author and their title which is shown in italics. Other details, like publisher, date and price, have normally been omitted in order to save space, and in the light of increased information technology which allows databases to be searched on-line. If readers have any difficulty tracking down the source of a particular item please contact the author.

# 4 Sustainability, equality, and pluriactivity: the suitability of the farm household as a vehicle for rural development

*Lyneth Davies, Nicholas Mack and Morag Mitchell*

## Introduction

This chapter takes a broad look at pluriactivity (or multiple job holding) in marginal farming households and discusses the role of these holdings in rural development policy. Findings from an analysis of farm household data from surveys in three peripheral regions — Northern Ireland; Scotland; and Wales — support a view of the household as an important locus of adaptation, formed from the individual objectives of household members negotiated within the collectivity of the household. These individual activities link farm households into the local rural economy and improve the economic resilience of peripheral agriculture.

However, the extent of household pluriactivity revealed by the surveys also begins to question traditional distinctions between farm and non-farm households. In particular, it calls into question whether marginal/pluriactive farms should continue to be a differentiated focus for agricultural as well as rural policy, or if it is more important to focus on sustaining peripheral rural communities as a whole.

## Competitive imperatives in agricultural and rural policy

In spite of the growth of pluriactivity — namely, multiple job holding additional to farming — and the rise of non-agricultural employment in the countryside, family farms are still recognised as integral to the social and ecological pattern of the countryside, inseparable from the nature and appearance of the landscape and many social, cultural and historic features of

*Progress in Rural Policy and Planning*, Volume Five. Edited by Andrew W. Gilg

rural areas. Increasingly, however, the role of small and medium-sized farms in relation to food production has been peripheralised by the dominant (globalising) pattern of agri-industrial development, and their future has been brought into question.

With the GATT agreement, an increasingly competitive climate will consolidate, in general, the market-led processes of agri-industrialisation and rationalisation, and in particular, the efficient distribution of food production to those areas of most comparative advantage. If market forces are to be the only criterion for deciding the future of agriculture, peripheral areas, such as Northern Ireland, Scotland and Wales, will move towards an industry concentrated on the most productive land, and on a few specialised products. Indeed, smaller farms (those less than 4 ESU (European Farm Business Size Units)) have been identified by the Department of Agriculture for Northern Ireland (1994), for example, as a structural weakness undermining agricultural competitiveness, leading to a call for measures to facilitate the occupational mobility of such farmers out of farming.

## The pluriactive strategy

Until the implementation of the 1992 reforms the CAP had provided little help to those farmers who wished to diversify into non-agricultural activities. The situation was further aggravated by post-war attitudes toward agriculture in UK planning policy, especially the anti-development ethos of the Town and Country Planning Act 1947. These attitudes protect agricultural land against development, and together with the limited funding of the Guidance section of the CAP, have inhibited the opportunities for linkages between agriculture, farm households and a wider rural economy.

Yet as Table 4.1 shows, pluriactivity was common to more than half of all households surveyed by the authors. This finding came from three separate but related projects under the 'Changing Farm Economies' project funded by the ESRC as part of the 'Joint Agriculture Environment Programme' (JAEP) between 1991 and 1993. These surveys of three peripheral regions in Northern Ireland, Scotland and Wales, which covered a total of 2100 farm households, also found that pluriactivity was strongly oriented towards off-farm employment. It is thus clear that many farm households are linked into local and indeed regional economies stretching beyond agriculture. Further, this 'household occupational mobility', rather than facilitating an exodus from farming, helps instead to sustain the marginal farm.

Thus a study of pluriactivity reveals the household as a locus of socio-economic adaption capable of sustaining the farm within a competitive agricultural climate as well as a collection of individuals other than the farm operator with interests apart from farming. Among a growing number of

**Table 4.1** *Distribution of pluriactivity among survey households (%)*

|  | Non pluriactive | Pluriactive | Off-farm pluriactive | On-farm pluriactive |
|---|---|---|---|---|
| Scotland | 41 | 59 | 49 | 18 |
| Wales | 34 | 66 | 53 | 27 |
| N. Ireland | 44 | 56 | 53 | 5 |

others, Wheelock and Oughton (1994) suggest that it is a mistake to base rural policy on the individual farm operator, but rather, to target it onto households or families. Equally, it may have been shortsighted to have focused agricultural policy for so long on the priority of land and farmer alone. What does this wider, household perspective mean in practice?

Households comprise a number of individuals linked by quite fundamental biological relations built on gender, lifecycle, conjugal and kinship ties. This means that there is an interest in acting together. Overlaid on these are socio-cultural factors ranging from long-held beliefs and ideologies to more immediate occupational status, affecting the profile of labour, decision making and resource consumption within the household.

Wheelock and Oughton (1994) introduce a cooperative conflict model of 'bargaining' between household members to illustrate a process by which individual members negotiate their contributions to and entitlements from household activities. These can change in relation to events in the social or economic environment affecting the household, and the perceived roles, status, opportunities or constraints of each individual. In terms of pluriactivity Fuller and Bollman (1992) suggest two paradigms: first, the displacement of labour from the farm by increasing technology and capitalisation, and second, changes in the socio-cultural environment of the household due to wider social change.

In the data presented in Table 4.2 different members demonstrate different levels of contribution to the general occurrence of household pluriactivity. The analysis of the farm household surveys we conducted around this feature indicates a diversifying set of individual agendas. Spouses in our three surveys consistently represented an important source of household pluriactivity. They were more often in possession of further, wider-ranging qualifications than farmers, and indeed, as Table 4.3 shows, the most common form of work was professional/technical. However, spouses were also often dependent upon the public sector for employment, as teachers or nurses, or in government administration. At the time of the survey at least, few were employed as professionals in the private sector and few women were self-employed.

Alongside this more widespread social change is a general weakening in the particular ideological certainty surrounding farming. As Marsden *et al.*

Table 4.2   *Pluriactivity in household members by region*

| Region | Pluriactivity | Household members | | |
| --- | --- | --- | --- | --- |
| | | Farmer | Spouse* | Sons/Daughters** |
| | % Pluriactive | 28 | 33 | 61 |
| Scotland | % PA off-farm | 67 | 82 | 91 |
| | % PA on-farm | 41 | 22 | 10 |
| | % Pluriactive | 31 | 44 | 41 |
| Wales | % PA off-farm | 73 | 65 | 92 |
| | % PA on-farm | 37 | 50 | 16 |
| | % Pluriactive | 25 | 29 | 50 |
| N. Ireland | % PA off-farm | 87 | 91 | 97 |
| | % PA on-farm | 13 | 9 | 3 |

\* Figures given are percentages of *households with spouses*, not all households
\*\* Figures given are percentages of *households with offspring*, not of all households or all offspring

Table 4.3   *Most common employment sectors for farmers and spouses in off-farm work*

| Region | Farmers | Spouses |
| --- | --- | --- |
| Scotland | 1. (42%) Production work<br>2. (18%) Agricultural Industry<br>3. (16%) Administration/managerial<br>Least — (<1%) clerical | 1. (41%) Professional/Technical<br>2. (17%) Service Sector<br>3. (16%) Clerical<br>Least — (<1%) agricultural industry |
| Wales | 1. (64%) Agricultural industry<br>2. (11%) Production work<br><br>3. (10%) Professional/Technical<br>Least — (<1%) clerical | 1. (32%) Professional/Technical<br>2. (18%) Agricultural industry/services<br>3. (16%) clerical<br>Least — (<1%) administrative/managerial |
| N. Ireland | 1. (50%) Production work<br>2. (16%) Agricultural industry<br>3. (11%) Professional/technical<br>Least — (2%) service sector | 1. (33%) Professional/technical<br>2. (27%) Production work<br>3. (26%) Clerical<br>Least — (<1%) Administrative/managerial |

(1993) have observed, the 1992 CAP reforms create not only pressures for structural change but also a crisis of identity, and a need for a re-assessment of the role and purpose by which the farming community legitimises its place in society. The 1993 GATT agreement adds to these pressures. However, the growth in 'environmentalism' in Western society, which places an increasing value on the welfare of the environment and appearance of the countryside, presents the farming community with an ideological niche other than productivity which can be colonised. This colonisation is already evident in farm union rhetoric regarding stewardship in the countryside (Collins, 1994), but this new ideology may not yet have been fully embraced by the farming community.

For example, farmers in our household surveys continue to demonstrate a commitment to agriculture, even where farming is economically marginal. They were also least likely to be pluriactive and our work has suggested that they were the first to withdraw from further activities as farm business size increased. Where they were pluriactive in off-farm employment they were commonly so in agriculturally related industries, or in activities which feature similar skills to those used in farming, as shown in Table 4.2.

Data available on the activities of offspring, meanwhile, including likely successors to the farm, varied between project surveys. Trends indicated that a significant proportion had off-farm jobs, did not typically feature in on-farm activities, and, increasingly, off-farm careers were found to be drawing offspring away from the farm household and farming altogether.

Changes in the social, economic and ideological environment of farm households might be said then to have encouraged a diversification in the economic and social objectives of individual household members, under-mining the centrality of the farm to household and individual welfare. Leaving aside the complex question of succession for the moment, this diversity can sustain a farming enterprise otherwise liable to displacement by larger, more centralised agriculture. Land thus remains employed in a marginal agriculture rather than being released for other economic purposes.

The increasing orientation of household members to the local non-agricultural economy, however, means that it is less easy to justify basing rural policy around the continuation of farming when the farmstead is increasingly becoming a homestead. The question therefore arises, should policy still recognise 'farming' in these marginal households as a case for specific policy support? Potential answers to this question are most important in the signals they provide for future successors and their willingness to take over a farm business. The concluding section of this chapter thus examines these perspectives in terms of wider rural policy objectives.

## Rural development and the pluriactive household

Post-productionist agricultural policy has to take into account the wider processes of rural social and economic restructuring, notably the demo-graphic changes brought about by changes in technology, global trade, and the role of the countryside. In some areas the process of in- and out-migration has contributed to an ageing population as young people have left in pursuit of careers elsewhere. The loss of young people may leave an area unable to compete regardless of actual factor endowments (Nevin, 1980). Equally, such processes often act to dilute and change traditional values associated with local and regional cultures, including the minority languages which underpin them. In this context, farm and non-farm households are inexorably linked in the challenges facing many rural areas to compete and survive.

Statements from the erstwhile European Community recognise this, seeing the aims of rural development as including:

> . . . the unavoidable adjustment of farming to actual circumstances in the market and the implications of this adjustment not only for farmers . . . but also for the rural economy in general (COM(88)501 final).

There is thus a need to evaluate if post-production policies contribute toward this wider picture, and the extent to which agricultural policy conflicts with or complements rural policy.

In a keynote address to the European Rural Development Conference, Bryden (1992) has highlighted the growing concern over subregional uneven development and centralised policy initiatives, which have benefited some areas and population groups more than others. He emphasised the need for locally initiated programmes in tackling the specific problems and opportunities of different subregions. In implementing such programmes, the European Union has placed an emphasis on indigenous human resources and a diversification of the rural economy to reduce the dependence on agriculture. In particular, it has highlighted non-agricultural developments such as Information Technology and Rural Tourism, and the maintenance of the rural countryside.

A notable example of this local level approach is provided by the EC LEADER Programme (Links bEtween Actions for the DEvelopment of the Rural Economy). LEADER is a community development initiative consisting of over 200 action groups in Objective 1 and Objective 5b regions. It aims to establish a network of local development groups which can employ a substantial degree of flexibility in implementing development objectives and in finding innovative solutions. Within such a framework, farm households are not specifically distinguished from non-farm ones, but constitute part of the 'innovative solutions' being sought. Such solutions, however, are subject to a different set of political, social, and hence ideological influences defining the role of land and farming in the locality.

For example, Antur Teifi is an Enterprise Agency operating a LEADER programme in the Teifi valley of West Wales. It actively seeks the 'socialisation' of farm households, drawing them into the wider rural 'economic development community', In doing so, the problem of a separative farming ideology is reduced through considering the household rather than just the farmer.

*Agriculture: the issue of 'functional marginality'*

The farming community is becoming inexorably modified into two groups: first, a smaller body of professional economically viable farmers who will inherit the ideology of production, and second, a larger mixed body of 'functionally marginal family farmers' left with a weaker — because less certain — 'stewardship' role. If this second group is to become the norm, and is to be reproduced via traditional family succession, a clear ideological focus or economic rationale must be defined, probably centred around an environmental role.

Thus the sustainability of agriculture — outside from the arena of the professional farmer — may well depend upon the activity of building more intimate relationships with the environment. In marginal areas, ESA management schemes, diversification and extensification are already blending agriculture into the local ecology, but in so doing have made farming increasingly unprofitable, uncompetitive, and grant dependent. Such policies may therefore undermine policies aimed at developing the wider rural economy. However, advocates of non-conventional forms of agriculture point out that such 'functionally marginal farms' by pursuing less intensive forms of agriculture may insure not only themselves but also agriculture as a whole — at both the regional and local level — against the dangers of over-specialisation and increasing ecological instability (Altieri, 1989; Douglass, 1984) which may be inherent in 'professional' farming.

The ideological rationale, therefore, may be that of a 'functionally marginal' agricultural infrastructure based on less intensive forms of agriculture which could form an 'ecological buffer' around areas of less environmentally friendly farming. As Battie and Taylor (1989) note, the opportunity of pursuing tangible goals towards such a diverse agriculture within the prevailing system does imply separate policies within agriculture to achieve separate (agricultural) goals. We would go further, however, and suggest that a separate marginal farming policy and associated ideology cannot be solely agricultural but needs to bridge the gap between farming and the non-farming interests in local rural economies.

Promotion of less intensive forms of agriculture will not in itself result in rural revitalisation, but it is widely acknowledged that conservation and landscape may be important resources for development, underpinning tourism, marketing and promotion of an area in attracting inward investment. Explicitly 'agri-ecological' policies, meanwhile, need the benefits of alternative sources of income if they are not to be solely grant dependent. This means additionally setting out to link diverse agriculture as a land use to the wider social and cultural fabric of marginal rural areas. Alternative land use must be part of a more common area based approach to economic development integrating agricultural and non-agricultural interests.

As Lampkin (1993) has reported, conversion costs associated with the transition to organic farming may be of the order of £150–200 per hectare per year for cereal and arable farms, £100 per hectare per year for specialist dairy farms, and £50–100 per hectare per year for livestock farms. These costs include the cost of training, information gathering, and the loss of income during the transition process.

Quite apart from the role of pluriactivity in 'buffering' against these transition difficulties, there s evidence to suggest that pluriactivity already has some (perhaps unintentional) benefits on local species diversity and the 'look' of the countryside, particularly where this includes *both* on-farm enterprises and off-farm work (Ellis and Heal, 1993). While this diversity may be an unintended side-effect of pluriactivity, perhaps the result of an associated reduction in management intensity, it serves to illustrate the potential.

Measures to encourage pluriactivity and the maintenance of farming within the 'distributed strategies' (Wheelock and Oughton, 1994) of farm households are thus important policy tools in formulating policy solutions which can sustain marginal farming. A diverse agriculture thus means explicitly recognising pluriactivity, not as a means out of farming but as a means of sustaining an integrated rural agriculture by providing other household members with a link between the farm household and the wider rural economy.

Perhaps the growth of environmentalism is thus a socio-cultural change which allows one or more members of the household to continue to farm, but if so it must be with the support of other household members in non-agricultural activities. A distinct policy signal from within the rural policy milieu is thus needed which recognises and builds on this interdependency. Such a policy signal will for the first time need to recognise the importance of people as well as land.

# References

Altieri, M.A., 1989, 'Agroecology: a new research and development paradigm for world agriculture', *Agriculture, Ecosystems and Environment*, 27(1–4): 37–46

Battie, S.S. and Taylor, D.B., 1989, 'Widespread adoption of non-conventional agriculture: profitability and impacts', *American Journal of Alternative Agriculture*, 4: 128–34

Bryden, J., 1992, 'Rural change and rural policy in the European Community', *Quarterly Journal of the Community Programme LEADER*, Winter 1992–93(2): 7–10

Collins, N., 1994, 'Agricultural policy networks in Northern Ireland and the Republic of Ireland', *Journal of Political Science* (forthcoming)

Department of Agriculture for Northern Ireland, 1994, *The State of Agriculture in Northern Ireland: Northern Ireland Structural Funds Plan 1994*, HMSO, London

Douglass, G.K., 1984, 'The meanings of sustainable agriculture', in Douglass, G. (ed.), *Agricultural Sustainability in a Changing World Order*, Westview Press, Boulder, Colorado, pp. 3–29

Ellis, N.E. and Heal, O.W., 1993, *Differences in farmland ecology associated with pluriactivity: Scottish 'Joint Agriculture and Environment Programme (JAEP): Report to the ESRC*, Institute of Terrestrial Ecology, Edinburgh

Fuller, A. and Bollman, R., 1992, 'Pluriactivity Among Farm Families: Some West European, US and Canadian Comparisons', in Bowler, I.R., Bryant, C.R. and Nellis, M.D. (eds), *Contemporary Rural Systems in Transition*, Vol. 2, CAB International, Wallingford, pp. 120–32

Lampkin, N., 1993, 'Economic aspects of the transition to more sustainable farming systems in the context of the reform of the Common Agricultural Policy', Paper presented at the Agricultural Economics Society Conference, *Problems and Prospects for a Sustainable Rural Economy*, London, December

Marsden, T., Lowe, P. and Murdoch, J., 1993, *Constructing the Countryside*, Academic Press, London

Nevin, E.T., 1980, 'Regional policy', in El-Agran (ed.), *Economics of the European Community*, 3rd edition, Phillip Allen, Oxford, pp. 232–255

Wheelock, J. and Oughton, E., 1994, 'The household as a focus for comparative research', Working Paper, 4, Centre for Rural Economy, University of Newcastle upon Tyne

# Section III:
# Europe


edited by
*Philip Lowe and Jonathan Murdoch*

# Introduction

*Philip Lowe and Jonathan Murdoch*

In the 1990s the high ambitions of the European pioneers have been sorely tested. The assertion of European ideals above those of 'nation' has become increasingly harder, as economic depression, political uncertainty and cultural nationalism have provoked reactions against a common European community. Nevertheless, it is generally recognised that the European Union has a momentum of its own, and many national groups and organisations are now willing to think as 'Europeans'. Thus, in the environmental field, where it is generally recognised that the problems know no frontiers, a coordinated trans-national response seems appropriate.

One of the chapters in this section, Rosie Simpson's discussion of sustainable tourism policies, is illustrative of this trend. It is increasingly recognised that tourism, in itself a trans-national activity, can be responsible for pollution, the consumption of scarce resources, the disruption of local communities and damage to wildlife. In response to these types of issues, and the absence of national policies, the Federation of Nature and National Parks of Europe has established a Green Tourism Working Group in order to examine how tourism might be made more sustainable. As Simpson explains, the Working Group recognised a need for more strategic action and the requirement for a much more developed policy framework within which such action might be directed. Such a framework would, of necessity, be quite complex, for it is also argued that a partnership approach, linking local communities, the tourism sector and managers of parks and natural resources, would be the most suitable means of achieving the objectives of sustainable tourism.

While the problems associated with environmental change may be forcing policy makers to think trans-nationally, the pressures created by expansion of the European Union are also pushing the issue of 'cohesion' firmly onto the European policy agenda. Thus one of the most notable policy initiatives at the European level has been the strengthening of structural policy in recent years. The aim of structural policy is to permit 'lagging' countries and regions to be brought up to the level of the more advanced. However, the chapter by Saraceno challenges some of the assumptions that lie behind such a policy, particularly as they relate to rural areas. She questions the ways in

*Progress in Rural Policy and Planning*, Volume Five. Edited by Andrew W. Gilg
© 1995 Editors and contributors. Published 1995 by John Wiley & Sons Ltd.

which spatial hierarchies are presently arranged — such as core–periphery, developed–underdeveloped, marginal–central, etc. — and argues that the important thing is not to predetermine development paths. Questions are raised, therefore, about the right mix of policy intervention/direction and indigenous growth. Can cohesion be 'masterminded' from the 'centre' in quite the ways expected by European policy makers? Or is structural policy simply a palliative designed to cushion certain areas against the worst excesses of the Single European Market?

In our review we argue that if the latter is the case then present policies are clearly inadequate. We show that structural policy is entering a new stage with new countries and regions set to benefit. We provide some of the details of the new Objectives and question whether resources are being concentrated sufficiently to yield real and lasting benefit to the 'lagging' regions. Meanwhile, other measures to achieve economic growth and competitiveness were presented in the Delors White Paper. This discussion document considered the direction the European economy should take in the medium and long term, particularly in relation to employment. While there are doubts about the effectiveness of many of its proposals, the White Paper makes some interesting suggestions about taxation policies, particularly the substitution of taxes on natural resource use for taxes on employment. This just might signal a profound shift in fiscal incentives away from energy and resources intensity towards use of labour. The Delors proposals were a notable contribution to general environmental policies in a year when the political momentum behind the 'greening' of the Union seemed to be slowing.

Likewise, in the area of agricultural policy the reforms associated with the MacSharry proposals appeared to have taken the political pressure off the agricultural sector. Even the GATT deal ultimately left many of the core policies intact. However, voices are already being heard, claiming that the reforms have not gone far enough and before long the old problems will soon be raised anew. The idea of 'Europe' seems as a result of these mixed fortunes to be quite tarnished. The grand hopes originally associated with the European project have been replaced by procrastination and uncertainty. With the advent of new, and relatively untested, policies, as, for instance, in the environmental field, the promise of Europe might seem to be capable of renewal. However, the mature policies still in existence, such as under the Common Agricultural Policy, do not inspire great hopes for the future of the Union.

# 5 European review 1993/4
## Philip Lowe and Jonathan Murdoch

## Introduction

The year began under a cloud of uncertainty and apprehension in Europe. Economic recession, rising unemployment and war in the former Yugoslavia all gave rise to fresh doubts over the robustness of the European project. Such doubts were further fuelled by the turbulent ratification processes in the member states on the Treaty of the European Union, which came into force on 1 November 1993. Likewise, the turmoil on the foreign exchange markets during the summer of 1993 and the partial breakdown of the European Exchange Rate Mechanism (ERM) under the speculators' assault seemed for a while to imperil the second phase of economic and monetary union, which began on 1 January 1994. However, despite the disruption to the ERM most member states indicated in the wake of the crisis that a coordinated approach to economic and monetary union was still their preferred option, with France and Germany in particular keen to press forward, although the timescale for full economic union is likely to be prolonged.

Not only did the Maastricht Treaty prove to be an anti-climax but it also left behind a legacy of popular distrust and nationalistic feelings. Beset by declining economies and rising unemployment, European governments became more than usually defensive of their national economic interests and inclined to backpedal on European integration. Industrial output in the European Union fell by 3.5 per cent in 1993, the worst figure since 1975, and unemployment rose to 11 per cent.

In an effort to regain the initiative, the European Commission promoted an EU-wide growth strategy during 1993–4 whose elements were drawn together in Jacques Delors's White Paper on Growth, Competitiveness and Employment (see the specific section on this, on pp. 105–7). In keeping with the prevailing orthodoxies of sound public finance, the emphasis was less on short-term job creation and the addressing of cyclical problems and more on labour market measures and the switching of public expenditure towards infrastructure and other capital investments seen to have a long-term effect on the supply potential of the EU.

The Commission has also had to respond to the incorporation of the

*Progress in Rural Policy and Planning*, Volume Five. Edited by Andrew W. Gilg
© 1995 Editors and contributors. Published 1995 by John Wiley & Sons Ltd.

principle of subsidiarity into the Maastricht Treaty (see PIRPAP, Volume Four, p. 210) and the pressures from some member states to reclaim areas of competence. In the prevailing deregulatory mood this has led to a wide-ranging review of existing EU legislation and a distinct slowing down of new legislative initiatives (see the specific section on the 'Application of the subsidiarity principle' on p. 107).

The completion of the Maastricht Treaty paved the way for a new phase in the enlargement of the European Union. Involving the EFTA (European Free Trade Area) countries, this promised to be an easier phase than either the previous one that brought in Greece, Spain and Portugal (all of which needed fairly elaborate transitional arrangements) or the next major phase of enlargement, to the East. The coming into force of the European Economic Area in January 1994 incorporating all the EFTA economies except Switzerland meant that already the candidate countries had accepted the body of EU legislation governing the Single European Market. As relatively wealthy countries with strong democratic traditions, high environmental and welfare standards and staunch commitments to both free trade and strong agricultural and regional supports, the prospect of their entry was generally welcomed across the European Union. However, public opinion within the EFTA countries themselves was more ambivalent about joining the EU, and Iceland and Liechtenstein decided not to seek membership.

Switzerland's application was put on hold following the rejection of the European Economic Area in a referendum in December 1992. Accession negotiations were held with the other states and were concluded politically in March 1994, but in a referendum in November 1994, 52 per cent of the Norwegian people who voted cast their votes against membership. It was the second time in 22 years that the Norwegian people had voted against joining the EU. Opposition was strongest in the north of the country among fishing and farming communities concerned about the impact on their livelihoods and autonomy of the Common Fisheries Policy and the Common Agricultural Policy. Norway thus stands alongside Sweden as a very prosperous country on the edge of the European Union whose people feel they have little to gain economically from membership but too much to lose.

However, earlier referenda in Austria, Sweden and Finland had accepted entry and these countries will join the EU at the beginning of 1995. The terms and consequences of this enlargement are considered below in the specific section on the 'Enlargement of the European Union' (pp. 107–9).

There remains a queue of other applicants from Central and Eastern Europe. No fixed timetable has been set for future enlargement as it is dependent upon the 1996 Intergovernmental Conference successfully reforming the EU's institutional structures and regional policies, to enable it to cope politically and financially with even more members.

Thus the year for the European Union was marked by increasing problems of integration and a tentative expansion. Inevitably these larger developments have influenced the development of policy in those fields that constitute rural policy. The overriding preoccupation with stimulating growth and the general deregulatory mood have not been auspicious for the development of environmental policy and there has been a distinct loss of momentum in this field. It is expected that the acceding states with their environmental concerns will help revive that momentum. However, it is in the environmental field where the principle of subsidiarity is being most thoroughly applied.

European regional policy, in contrast, has been given a boost by the growth initiative. A new six-year programme of ECU 140 billion of Structural Funds was agreed in 1993, focused on training and capital investment in deprived and declining regions. The accession negotiations introduced a new objective for structural funds to cover the very sparsely populated northern regions of the Nordic countries. The Cohesion Fund also came onstream with ECU 14 billion over six years for transport and environmental projects in Portugal, Spain, Ireland and Greece.

Although agricultural policy had seemingly been the subject of anguished negotiation, first around the MacSharry proposals and latterly within the GATT Uruguay round, in many respects this policy area conducted 'business as usual'. The pressures for economic liberalisation, deregulation and subsidiarity barely seemed to trouble the CAP. Even where reform was implemented, i.e. in the cereals sector, the result was increased levels of bureaucracy and financial support. During the year, it became a little clearer that 'reform' of the CAP might not have addressed the agriculture sector's key problems.

The rest of this chapter considers some of the matters discussed in this introduction, in more detail, and specifically: The Delors White Paper on 'Growth, Competitiveness and Employment'; application of the subsidiarity principle; enlargement of the European Union; environmental policy; agricultural policy; and rural development.

## The Delors White Paper on 'Growth, Competitiveness and Employment'

This document (European Commission (93) 700 Final) must be seen in many respects as Jacques Delors's swansong. Having guided the European Union through the completion of the Single European Market and the Maastricht Treaty, Delors stepped down from the post of President at the end of 1994. The White Paper was typically upbeat and forward-looking. It was presented to the Brussels European Council in December 1993 and was the main subject of the Corfu summit in June 1994.

It is not a White Paper in the normal sense of a set of proposals for directives which would make their way through the EU's legislative process. Instead it is a discussion paper on the direction the European economy should take in the medium and long term. It sets out a strategy to 'lay the foundations for sound and lasting economic growth', with a target of at least 3 per cent growth per annum and a halving of the present rate of unemployment in the EU by the year 2000 with the creation of 15 million new jobs. A central concern of the White Paper is the 'employment intensity' (i.e. the job content per unit of output) which is markedly lower for the dynamic sectors of the European economy that for that of its main industrial competitors.

Indeed, as presently structured, the European economy is probably unable to grow fast enough to offset the secular growth in the workforce and thus stabilise unemployment levels.

The White Paper therefore looks to a number of ways of enhancing the capacity for labour-intensive economic growth, including a combination of labour market measures to improve competitiveness and reduce working time and macro-economic measures to expand demand, particularly for goods and services with a high employment intensity. Much of this would be in personal services (such as child care, home help, youth work and leisure) but the White Paper also refers to 'renovation of old housing', 'local public transport services' and environmental protection (including 'maintenance of natural areas and public areas', 'water purification', 'monitoring of quality standards' and 'energy-saving equipment'). It also suggests some really large new demand sectors, notably a huge expansion in transport, information and energy networks.

It is the latter sectors which have attracted most attention, particularly the proposal that the EU should fund the development of strategic trans-European networks both within the EU and towards the potentially important markets of Central and Eastern Europe. The Corfu summit agreed 11 major transport projects (mainly high-speed rail schemes and motorways) and gave priority status to eight energy projects (mainly for gas distribution).

While doubt must remain over whether demand in these sectors could be expanded sufficiently to yield the 3 per cent growth target for the overall economy, that target itself sits uneasily with the EU's stated commitment to environmental protection. Such a growth rate would inevitably mean more development on greenfield sites, more emission-causing transport, and more use of non-renewable resources. The conclusions of the White Paper do acknowledge that 'A basic challenge of a new development model is to rectify the current misconception that there is a negative relationship between, on the one hand, environmental protection and general quality of life and, on the other hand, economic development'. But it has few tangible suggestions on how environmentally sustainable growth might be achieved. The most interesting suggestion relates to tax reform through a shift in the burden of taxes away from labour, since such taxes are thought to penalise employment. Given that member states have labour taxes that make up 60–90 per cent of state revenue and very few taxes on non-renewable energy, raw materials and other ecological factors, a readjustment could make the economy more labour-intensive and less energy- and resource-intensive, with benefit to the environment and job creation. However, the one practical proposal of this kind already on the table — a $CO_2$ tax — has remained deadlocked in the Council of Ministers.

## Application of the subsidiarity principle

The Treaty on European Union (see PIRPAP, Volume Four, pp. 209–10), which entered into force on 1 November 1993, introduced the principle of subsidiarity as follows:

> In areas which do not fall within its exclusive competence, the Community shall take action . . . only if and in so far as the objectives of the proposed action cannot be sufficiently achieved by the Member States and can therefore, by reason of the scale or effects of the proposed action, be better achieved by the Community (Article 3b).

At the Edinburgh European Council in December 1992 the European Commission gave three undertakings with regard to the application of the subsidiarity principle: that a justification would be included in all new legislative proposals; that it would withdraw or revise certain pending proposals; and that it would review existing legislation. A year later it reported on progress to the European Council in Brussels in December 1993. Detailed examination of the subsidiarity principle led to a reduction in the number of new legislative proposals put forward by the Commission in 1993 compared with previous years. Some 150 proposals, which were technically outdated or politically obsolete, were withdrawn. The Commission also began the enormous task of reviewing existing Community legislation giving priority to old legislation but also prodded by submissions from the French and British governments jointly and from the German government identifying hit-lists of EU measures to be amended or abolished. By December 1993, some 50 000 pages of legislation had been reviewed and the Commission anticipated scrapping up to a quarter of it as outdated and to consolidate and streamline much of the rest. Among the fields most affected are product standards and the environment. (See pp. 109–11 below.)

## Enlargement of the European Union

The year 1993/4 also saw the negotiations which led to Austria, Finland and Sweden becoming members of the European Union (the formal date of accession is 1 January 1995). The degree to which these countries were already fully integrated into the Single European Market, plus their small size and relative wealth, made for comparatively straightforward negotiations on the terms of their entry, which were confirmed in parliamentary ratifications and national referenda conducted through the summer and autumn of 1994.

The expansion to 15 member states will increase the population of the EU by 21.6 million to a total of 367.7 million. The new members have a higher than average GDP per head than the EU average, and Sweden and Austria will be significant net contributors to the EU's budget. The Union will also benefit from the geographical position of the new members: Austria lies at the heart of Europe with trade links to Eastern Europe; Sweden has strong trading links with the Baltic States; and Finland has long experience of trading with the Commonwealth of Independent States. The three new members also bring with them a strong tradition of openness in government, social equality, and high environmental, health and safety standards.

These higher standards and how they would affect the free circulation of goods were one of the key issues addressed in the entry negotiations. The compromise agreed was that the applicant countries could maintain their national standards for a transitional period of four years during which time a review of EU provisions would take place, with the new member states fully involved. The net effect will be to strengthen significantly the bloc of environmentally committed northern member states within the Union.

One particular issue in the negotiations over Austria's accession has served to inject a greater environmental orientation into EU transport policy. For the Austrians heavy lorry transit through their country was the most politically sensitive matter in their entry negotiations. The environmental threat to the Alpine passes had already led to a bilateral agreement between Austria and the EU in which the number of vehicles was controlled by means of an 'ecopoint' system of transit licences. Each member state was allocated a certain number of 'ecopoints' to cover the passage of its lorries across Austria. In the entry negotiations, the commitment was made to continue with this system in the medium term if the Union was unable to find another way to reduce pollution through the Alps.

It was agreed to maintain the objective of reducing exhaust pollution by 60 per cent by 2003 and that the Commission should present proposals for EU-wide action to solve environmental problems resulting from road transport. This was vital due to the Swiss referendum on 20 February 1994 banning road transit through the Alps by 2004, with the possible effect of increasing the volume of transit traffic through Austria and France.

Aid for remote rural regions and agriculture were other key negotiating issues. In the applicant countries, especially the Nordic ones, a strategic objective of regional policy has been to maintain the population in remote northern and border regions. The Union recognised the handicaps facing the Nordic candidates, namely a short growing season and the isolation of border regions, and agreed to create a new type of regional policy ('Objective 6') for regions with a population density of less than 8 inhabitants per square km. In the agricultural sector farming has been well protected and supported. To cushion farmers against lower prices within the EU the establishment of degressive national aids to compensate for income loss for a five-year transitional period was agreed. In addition, farms in unfavourable and mountainous geographical locations will be eligible for support under Community schemes. The creation of 'Nordic support' involving long-term compensatory national aids for farms situated north of 62° latitude offered an additional safety net. More details are provided in the section on pp. 119–20.

## Environmental policy

During the 1980s the Environment Directorate (DGXI) of the European Commission could be portrayed as a legislative machine, pumping out draft directives. Amid the general optimism surrounding the European project, the environment emerged as central to the process of European political integration and was the fastest-growing field of EU legislation.

Certainly this has now changed, and the main preoccupations within the Commission are stimulating economic growth; pushing for deregulation; and reacting to calls for subsidiarity from member states. The drop in electoral support for the Greens in a number of countries has taken the pressure off national governments, and at the EU level environmental policy has visibly lost momentum, notably since the departure of the flamboyant Environment Commissioner, Carlo Ripa di Meana, in 1992.

Commission officials, though, are apt to present DGXI's diminished legislative output as a maturing of European environmental policy. They emphasise the steps taken to integrate environmental considerations into other aspects of EU policy and to establish formal consultative and monitoring structures for environmental policy. Undoubtedly, though, 1993–4 was a year of few tangible achievements and some significant setbacks.

*Policy achievement*

Much of the work of DGXI was in revising and updating existing legislation. A draft directive covering various technical revisions to the Environmental Impact Directive was published. More significantly, early in 1994, the Commission adopted a draft covering the fundamental revision of the so-called Seveso Directive. A Commission study had revealed that approximately 90 per cent of 130 serious accidents recorded in industrial undertakings during the Directive's ten-year period of application could be attributed to management failings, organisational shortcomings, inadequate training and human error. The Commission therefore proposed detailed emergency planning and a precise definition of in-house responsibilities; detailed safety reports; the construction of dangerous plants away from densely populated regions and vulnerable natural areas; EU-wide standardisation of control systems; removal of the distinction between production and storage in the case of safety measures; and greater public information, for example through access to safety reports on plants covered by the Directive.

The year saw considerable effort in reviewing the EU's pollution control policy. This involved the revision, simplification and consolidation of a lot of existing legislation on air and water. In future, water protection will be based on directives on drinking water quality; the ecological quality of surface water; the quality of bathing water; freshwater management and ground-water protection; urban waste water; and protection of waters against pollution by nitrates from agricultural sources. As far as air quality is concerned there is no definition of objectives or harmonised monitoring criteria for Community air quality standards, although there is a sizeable volume of legislation on source emissions. The Commission plans to work on the definition of objectives before embarking on its review of existing legislation.

With the application of the subsidiarity principle, there is some evidence of a partial move away from prescribing uniform at-source emission limits and towards establishing equivalent national procedures for determining and achieving environmental quality objectives. Environmental groups have been critical of such an approach, arguing that the national flexibility allowed will be at the expense of the environment; will lead to looser and geographically variable standards; and will lack the progressive element that emission limits linked to the concept of 'best available technology' have provided.

In September 1993 the Commission proposed an integrated approach to the control of air, water and land pollution from large industrial plants in a draft directive on Integrated Pollution Prevention and Control. New industries would be expected to introduce the latest technology to reduce pollution, while older plants would have until 2005 to clean up and comply. The Commission proposed a coordinated licensing procedure based on admissible levels of pollutants for each industrial sector, member states setting their own emission limits within EC margins.

The Commission proposal for a directive on the ecological quality of water applied to any surface water not yet subject to a specific directive. The definition of quality is linked to purity and oxygenation. The proposal stipulates that the member states should: define quality objectives for all surface waters; establish a system to control water quality and an inventory of pollution sources; and prepare and implement a series of integrated programmes to enhance the quality of water. Immediate doubts have been raised about its likely effectiveness because it leaves it up to member states to decide how fast water quality should be improved. But there is little question that it will require them to devote more effort to monitoring and measuring the status of aquatic ecosystems and the causes of damage to them.

Under the proposal to amend the 1976 Directive on bathing water quality, the existing rules would be streamlined and clarified, and focused on the parameters of public health significance. Member states would also be given considerable latitude to remedy pollution of designated waters without incurring legal sanctions from the Commission. But the revised rules would almost certainly increase the number of bathing waters failing to comply with the Directive — in particular, the controversial standard for enteroviruses and a new standard for faecal streptococci.

Other draft directives issued during 1993–4 include the framework directive on air quality and one on the marketing of biocides. The European Commission also published a Green Paper on civil liability for environmental damage. Regulations were finalised during the year for the Eco-auditing and Eco-labelling schemes — these are both voluntary schemes intended to promote high standards among European firms.

In 1993 a non-structural fund, LIFE, the Financial Instrument for the Environment, was established. With a budget of some ECU 400 million for the period 1993–5, this is now the main source of EU funding exclusively for environmental initiatives. A first batch of projects under this scheme totalling ECU 65 million was adopted in 1993 for various initiatives promoting the use of so-called clean technologies, waste management, restoration of contaminated sites, the integration of the environment and the urban environment.

*Setbacks*

The year 1993–4 saw some significant setbacks for EU environmental policy in a number of fields including the urban environment, coastal management, climate change and wildlife protection. These illustrate not only the increasing recalcitrance of some member states but also the growing influence of interests opposed to strong European rules for environmental protection and a decline in the standing of DGXI within the Commission.

Plans for legislation to follow up the Green Paper on the Urban Environment (approved by the Commission in June 1990) were shelved. The Green Paper was an ambitious document that had stimulated considerable interest among local groups, politicians and planners across the EU. It sought to give a clear spatial focus to the EU's largely sectoral efforts at environmental regulation, including, for example, far-reaching proposals for traffic management and the integration of transport and land-use planning — in order to curb air pollution and improve urban conditions. The possibility that any of this might find its way into EU regulations was blocked, however, by the implacable opposition of the German Länder who, concerned to protect their constitutional position, objected that as the Federal Government had no legal competence in urban planning then neither should the European Union.

There seems to be no prospect of DGXI producing a parallel document on the rural environment.

Another field of spatial environmental policy where high expectations are likely to be frustrated is that of coastal management. A strategy for integrated management of Europe's coastal zones was originally requested by the Council of (Environment) Ministers in 1992. In March 1994 the Environment Council passed a resolution requesting the Commission for the third time to present proposals. The difficulty, it seems, arose from disagreements within the Commission, particularly between DGXI and the development-oriented DGXVI (Regional Policy). It now seems very unlikely that, if and when the strategy does appear, there will be proposals for new legislation or funds.

Back in 1990 the Environment Council made the commitment to stabilise $CO_2$ emissions at 1990 levels by the year 2000, but here again problems have arisen. In 1992 the European Commission drew up a Community Strategy to Improve Energy Efficiency, a central feature of which was a new energy/ carbon tax on fuels (determined partly by the energy content and partly by the carbon content). The tax was seen as the chief means to achieve the stabilisation of $CO_2$ emissions, and was promoted as such by the northern member states. The proposal, though, remained deadlocked in the Council of Ministers with southern member states concerned at the consequences for their industrial development and the UK government refusing to accept the principle of EU legislation on this subject. The UK government's opposition can be traced to its unpopular experience when it attempted to impose VAT on domestic gas and electricity bills in two stages between 1994 and 1995. The government had to cancel the second stage in 1995 after a defeat in Parliament in November 1994, and partly recouped the lost revenue by raising duty on petrol. Opposition also came from traditional industrial sectors and the oil companies. The prospects for an energy-carbon tax do not look good (the proposal was finally killed off at the Essen summit in December 1994) and it is not clear how the European Union will achieve its target emission reductions by the year 2000. Indeed, the disagreements over the tax have diminished the chances for a coordinated EU approach (including burden sharing between states) to meeting the Convention on Climate Change agreed at the Earth Summit. They have also dealt a blow to advocates of fiscal measures as a means to reflect environmental costs throughout the economy.

A final example relates to efforts to weaken EU wildlife protection. On 19 January 1994 the European Court of Justice ruled that, under Article 7(4) of the Birds Directive (79/409), hunting seasons must close to give complete protection to species during the period of pre-mating migration and that any variations in closing dates between species or between different parts of the same country must respect the principle of complete protection. The judgment caused uproar in French hunting circles and within 5 weeks, the

Commission (under pressure, it is thought, from its President, Jacques Delors) issued a draft directive removing the principle of complete protection by allowing member states more flexibility over 'birds with a good conservation status'. The proposal was given initial support by the Council of Ministers but must go before the European Parliament. It may not succeed but its introduction shows how vulnerable existing legislation can be in the present climate to a backlash from interests opposed to conservation.

*The integration of the environment into other EU policy fields*

In June 1993 the Commission adopted a number of internal procedures and measures for taking greater account of the environment when formulating its policies. Henceforth, any project liable to have an impact on the environment will have to be the subject of a strategic environmental impact assessment, and each year the Commission's 23 Directorate-Generals will provide a report on their activities with regard to the environment. Officials responsible for integration were appointed in each Directorate-General to coordinate this process.

The Commission made some progress in 1992–3 in beginning to integrate environmental concerns in the fields of transport, energy, industrial competitiveness and revision of the Structural Funds. In 1993–4 the focus moved to agriculture.

At an agri-environmental conference held in Gent in October 1994 Environment Commissioner Ioannis Paleokrassas emphasised what he saw as the shortcomings of the reformed Common Agricultural Policy from the standpoint of the environment. This policy 'is still very distantly linked to environmental policies' and even when this is the case, 'It is more often by omission than by commission', he declared. The agri-environmental programmes introduced under the reform are not themselves applied satisfactorily in most member states, he regretted, but he did pick out two notable exceptions, Germany and the United Kingdom. In addition to recommending more rigorous implementation of the provisions, the Environment Commissioner commented that (1) non-rotational set-aside for cereals must be compulsory to make this system more compatible with the environment; (2) farmers must be given the incentive of higher subsidies to use fewer fertilisers, pesticides and other inputs harmful to the environment; (3) it would be harmful for the environment for national governments to be given more responsibility in agricultural decisions because of the entrenched influence of national farming lobbies; and (4) zones with high ecological value should be created. As yet, there has been no formal response by the Agriculture Directorate, DGVI, to these criticisms and proposals

Meanwhile, some of the efforts to green regional and transport policies of the previous year seemed to come unstuck. For example, the new Cohesion

Fund, set up in the Maastricht Treaty to help the EU's four poorest countries — Greece, Ireland, Portugal and Spain — close the gap on their richer partners, was intended to make money available for transport infrastructure and environmental projects. Since it began operating in April 1993, environmental groups have been concerned that the eligible governments have been keener to use this money for transport expansion rather than environmental protection or sustainable development projects. Roughly two-thirds of the money has been spent in the transport sector, most on road projects. As for environmental spending, projects have concentrated on water supply, sewage treatment, waste disposal and pollution clean-ups. Some of this spending may actually damage the environment; for example, Greece is receiving ECU 67 million for a project, that environmental groups strongly oppose, to expand the Evinos dam which supplies water to Athens.

Likewise, European Structural Funds, though now meant to ensure that greater account is taken of the environment in regional economic development, continued to support large-scale, environmentally harmful projects in the continent's most economically backward regions. Examples include the trans-European transport networks; Spain's hydrological plan which makes excessive demands on water reserves in the north of the country by aiming to cover the whole territory with long-distance water pipes; outsize dams like that on the Acheloos River in Greece; and the large-scale afforestation of Southern Europe with foreign species.

The European Commission itself admitted that the environment 'seems to have been neglected' in a review of progress on the White Paper on Growth, Competitiveness and Employment prepared in advance of the Corfu summit held in June 1994.

*Consultative and monitoring structures*

Under the 5th Environmental Action Programme, the Commission is committed to the concepts of shared responsibility and partnership. Accordingly, it has helped to set up various consultative and cooperative monitoring structures. The General Consultative Forum on the Environment brings together leading industrialists, business executives and environmentalists three to four times per year to advise the Commission on how, at the European level, to achieve sustainable development. The Commission has also encouraged the creation of fora and partnerships bringing together industrial associations and environmental groups within different sectors.

An EU-Law Enforcement Network has also been set up. This brings together the national inspectorates responsible for the enforcement of EU environmental regulations within the member states. The aim is to encourage more exchange of experience between the Commission and the national authorities responsible for applying EU law. The hope must be that there is better feedback between the implementation and design of environmental regulations, and a transfer of expertise and knowledge between countries to improve on the patchy implementation of environmental directives.

*The European Environment Agency*

The long-delayed decision on the siting of the European Environment Agency (EEA) was eventually taken at the Brussels European Council meeting in October 1993, thereby allowing it to be established in 1994. A special EEA Task Force, though, had been working within the European Commission since 1990, when a Council Regulation had been passed to set up the Agency.

The purpose of the EEA is to provide reliable and comparable information at the European level to guide the development and assessment of environmental policy and to inform public debate. The Agency is to publish a report on the state of the environment every three years.

The EEA is required to make the maximum use of resources already existing in the member states, as much of its task will be in ensuring the standardised collection of data nationally. Information will be collected on the quality and sensitivity of the environment and the pressures it faces. In seeking to specify what data it will need the EEA is being guided in the first instance by the Fifth Community Action Programme on the Environment, 'Towards Sustainability' (PIRPAP, Volume Three, pp. 277–9). In its first years it will give priority to a number of areas including air quality, water quality, waste management, land use and natural resources. Within the next two years the Council of Ministers must decide on further tasks for the Agency, in particular with regard to monitoring the implementation of EU environmental legislation and the promotion of environmentally friendly products and technologies.

## Agricultural Policy

*GATT*

The year was significant for the achievement of a GATT deal on 15 December 1993. Subject to ratification by national governments (117 countries participated in the Uruguay Round) the agreement will come into effect on 1 July 1995 and run until 30 June 2001.

For agriculture the deal had implications in three main areas: domestic support; market access; and export subsidies. On the first of these, domestic support, the agreement entailed a reduction of 20 per cent over six years based on an Aggregate Measure of Support for the period 1986–8 (the AMS represents the difference between internal and world prices multiplied by the volume of production in this reference period). It is expected that the combination of reforms since that time, particularly those agreed in 1992, would ensure that the EU's total AMS would remain below the necessary figure throughout the six years covered by the GATT agreement.

In terms of market access it had been agreed that all border-protection measures would be converted into tariffs which would then be reduced by 36 per cent over the six-year period. In terms of export subsidies the volume was to be reduced by 21 per cent and budgetary expenditure on export subsidies by 36 per cent over the same period. There is some flexibility ('front-loading') allowed in the phasing of these cuts (a concession won by the EU). This allows 1991–2 to be used as a starting position rather than the base period 1986–90, which means the EU (and the USA and others) will be able to export more over the six-year period, thus helping the disposal of existing stocks. The EU reckoned that 'front-loading' would enable it to export an extra 8.1 million tonnes of cereals, 360 000 tonnes of beef, 250 000 tonnes of poultry and 100 000 tonnes of cheese. However, the agreed 21 per cent reduction from the base period 1986–90 must be achieved by the year 2000.

In general, the deal was clearly of great significance for world agriculture. Agra Europe (17 December 1993) summarised the major gains as:

(1) Agricultural trade will be subject to a set of binding international rules which in future will prevent the larger and richer countries from indulging in the damaging subsidisation of exports.
(2) There will be less scope for domestic governments to increase support for agriculture and then shift the consequences onto the international market.
(3) As a result of the limitations on exports and domestic support, the prices of agricultural products cannot now be increased by government manipulation.
(4) The initial freezing and eventual reduction of the level of export subsidisation and the modest opening of export markets combined with the scaling down of tariffs must lead to rising world prices.

*CAP reform*

While the GATT deal had far-reaching implications, the European Commission remained optimistic that the implementation of the Common Agricultural Policy (CAP) reform package, adopted in 1992, would be sufficient to reduce surpluses. The line pursued by the Commission clearly aimed to reassure the French government that the GATT deal would be contained within existing reforms and also to reassure third countries that the EU would curb surplus production by the late 1990s. At the core of CAP reform was the three-year programme of phased reductions in cereal prices, with compensation per hectare being paid directly to producers provided production was cut back by means of set-aside. By the end of 1993 the Commission was optimistic that the programme would prove effective for the following reasons:

(1) Arable farmers had reduced their use of inputs — not only due to set-aside but also, the Commission believed, because the compensatory 'area payments' discouraged farmers from maximising output on the 85 per cent of their arable land which remained in production.
(2) There had been an 'exceptional' uptake of set-aside, with almost 5 million ha taken out of production.
(3) As a result of set-aside, cereal production in 1993 was likely to be 20 million tonnes less than it would have been without reform.
(4) The lower price of EU-produced grain would mean an extra 12 million tonnes of EC grain being used in animal feed (Agra Europe, 29 October 1993).

The successful conclusion of the GATT negotiations and the seeming effectiveness of CAP reform led the Agriculture Commissioner, Rene Steichen, to argue (at the Oxford Farming Conference in January 1994) that there would be no need for further increases in set-aside or cuts in prices by the end of the current Uruguay Round agreement. With the announcement of the 1994 price package in January 1994 these issues were given further clarification as the Commission expressed confidence that the 1992 reforms would allow it to meet the cereal export reduction commitments under the GATT agreement. The decline in cereal prices (7.7 per cent in 1994) should, it was believed, dissuade farmers from maximising yield and outputs.

However, notwithstanding the Commission's optimism, it appeared by the end of 1993 that cereal production would be down by only 1 per cent over the previous year, raising doubts about the long-term effectiveness of the reform package (particularly as set-aside is likely to prove less, rather than more, effective as time goes on since average yields will tend to get higher on remaining productive land). However, in other key sectors, notably beef and sheep, output levels seemed to be broadly manageable and indicated to policy makers that no further changes would be necessary in the immediate future (although it was proposed that the reference year for special beef premium payments — 1992 — be changed as a result of the steep rise in such claims). The dairy sector, meanwhile, looked likely to remain in the straitjacket of quotas.

Many commentators became concerned that the achievement of the GATT deal, and its compatibility with the 1992 reforms, might lead farm ministers into a state of complacency. It was stressed by the Commission at the start of 1994 that slippage in the reform package should be avoided. However, the initial negotiations around the 1994 farm price review seemed to indicate that a sense of urgency had been lost. The Commission proposed that butter prices should be reduced by 5 per cent and milk quota by 2 per cent. Cereals and beef prices were governed by the reform package and most other prices were to remain frozen.

The Commission was concerned that savings on farm expenditure should be achieved in the price review, for budgetary problems still remained. In 1994 the guarantee section of the EAGGF, adopted in July 1993, was to amount to ECU 36 465 million (compared to ECU 34 052 million in 1993). This level exceeded the previous year, and was fixed at the same level as the so-called 'guideline', because the implementation of the CAP reform was likely to give rise to substantial expenditure. In the 1994 price negotiations the Commission stressed the need to agree a settlement which eased budgetary pressure. However, the ministers seemed more concerned with restoring the confidence of farmers and consolidating the reforms of previous years. Particularly contentious was the change in the reference year for the beef premium, but they also felt that the 5 per cent reduction in butter price was too tough and many had reservations about the cut in dairy quota.

After much procrastination (the price negotiations became entangled for a while with the German refusal to import British beef in the wake of fears over BSE and proposals to increase Italian and Greek dairy quotas) the price package was finally agreed in July, six months after the Commission's proposals were first tabled. The reference year for beef premium quota allocations was changed and it was agreed that the butter price should be reduced by 3 per cent (with a 2 per cent cut introduced in 1994/5). However, milk quotas were to remain unchanged for 1994/5 and 1995/6. Speaking after the meeting, René Steichen estimated the additional cost of ECU 81 million in 1994 and ECU 293 million in 1995.

While the haggling around the price package resembled that of previous years the reforms agreed in 1992 obviously constrained the amount of 'horse-trading'. Yet the incremental growth of agricultural funds, despite the limits imposed in the previous year by the 'guideline' (any expenditure exceeding the guideline can only be sanctioned by a 'jumbo' summit), continued to cause concern. The dairy support regime continued to be a drain on Union funds, costing, on average, around ECU 5 billion per year.

Moreover, it began to emerge that expenditure on area compensation and set-aside for cereals was likely to outweigh any savings in export restitutions and intervention. The Commission's figures showed that cereal expenditure under the new regime in 1994 was likely to be 100 per cent up on the last year of the old regime, 1992, and is projected to remain at this level until the end of the Uruguay Round period (2001) (Agra Europe, 15 April 1994). Thus agricultural spending threatens to continue surpassing the 'guideline' for the foreseeable future.

This was also evident from the Commission's proposed reform of the wine sector, aimed at eliminating wine lakes and improving quality, also outlined during 1994. It was suggested that wine production quotas, to the tune of 154 million hectolitres, should be introduced in 1995/6. The aim of the reform was to end the structural surplus through changes in levels of production rather than on storage and disposal of surplus stocks. However, the proposals, even if adopted in the form suggested by the Commission, would still entail the need for expenditure on the latter measures as it was expected that 20 million hectolitres of surplus wine per year would continue to be produced.

*Agriculture in the accession negotiations with Austria, Finland, Sweden and Norway*

Aside from the continuing struggle to contain the agricultural budget, the other main issue running through the year was the impending enlargement of the Union. Throughout 1993 and 1994 negotiations were conducted in order to clarify the transitional arrangements for agriculture in the wake of the impending accession of Austria, Finland, Norway and Sweden (in the end, the Norwegian people voted not to join). The Commission proposed that agricultural prices in the applicant countries should be aligned upon accession (scheduled to be 1 January 1995), with transitional aids paid directly to farmers. This proposal met with an angry response in each of the capitals. There were also problems on the appropriate beef and dairy quotas to be applied in the applicant states, with the latter seeking quotas greater than their existing production levels.

Accession is likely to affect agriculture in each of these countries differently. In Finland, agriculture is central to the economy and polity alike, thus the negotiation of CAP entry terms would be crucial to public perception of the Union itself. The country sought an alignment of prices but wanted a transition period. It was claimed by the Finnish government that the alignment of prices all in one go would leave farmers FM 2.6 billion in the red. The government also sought permanent structural measures, including the designation of the whole country as a less favoured area. In Sweden, on the other hand, it was argued that agricultural prices had already broadly come into alignment with those in the EU and the government was prepared for an immediate adaptation. The sticking point as far as the Swedes were concerned was quotas. They insisted that quota levels should reflect recent levels of production in meat, sugar and milk.

Austria also insisted upon a transition period and, like Norway and Finland, argued that as the EU was forcing the pace of the transition it should be prepared to pay compensatory aid. It sought a change in the Less Favoured Area criterion (i.e. that the altitude measure should be reduced from 800 metres to 500) in order to allow 80 per cent of its farms to qualify for such status. Norway was particularly concerned about the effects of accession upon its agriculture. The Norwegian government thus pressed for quotas that would allow an expansion of the suckler herd as an alternative to dairy production for some farmers. However, the country sought to gain from the designation of Objective 6 status (a new structural fund designation for Nordic areas above the 62nd Parallel and adjacent areas) of its four northernmost counties.

The applicant countries succeeded in winning a transition period for their overall budgetary contributions and permanent structural supports. However, the EU effectively achieved its demand for an immediate alignment of producer prices (with temporary compensatory support measures) and gave little ground on questions of quotas. Finland was set to receive ECU 457 million in compensation for directly aligning its prices, the money to be paid out over a four-year transitional period. Although it had been pushing for LFA status for the whole agricultural area it eventually settled for 85 per cent coverage. In addition, the new Objective 6 criterion would be applied in areas where population density is below 8 head of population per square kilometre. As in the Finnish case, Sweden was able to take advantage of Objective 6 status, particularly in the northern and central areas (around 60 per cent of the country). Austria too was set to receive extensive structural funds (under Objectives 2–5b and LFA designations) and seemed satisfied that sufficient transitional support had been granted despite some criticism over the levels of quota for both dairy and sugar.

## Rural development

### Changes in the structural programmes

The Delors II package (PIRPAP, Volume Four, pp. 211–12), approved at the Edinburgh summit in 1992, provided for a doubling of structural resources available to the four least prosperous Community countries up to 1991 under Objective 1 status and a major increase in funds under the other Objectives (Objective 1 covers regions whose development is lagging behind the rest of the Union; Objective 2 applies to regions seriously affected by industrial decline; Objective 3 aims to combat long-term unemployment; Objective 4 seeks to integrate young people into the occupational structure; Objective 5a seeks to modernise agricultural structures; and Objective 5b promotes development in rural areas). The next programming period will run from 1 January 1994 to 31 December 1999.

Following the Edinburgh meeting, the Commission presented its proposals on how the 'new' Structural Funds should be adopted and administered. The Commission's changes were by no means radical and really concerned only eligibility criteria, programming periods and administrative procedures. At a

Special Council meeting in July 1993 the new Regulations were approved and the list of Objective 1 regions was adopted, including areas in France, Belgium and the Netherlands for the first time. By the end of 1993, the Commission had agreed the distribution of the Structural Fund budget between Objectives and member states.

During the year, assessments of the earlier programme began to appear. Rural development issues were most notably addressed under Objective 5a and 5b. Under Objective 5a the countries to benefit most from the measures aimed at adapting agricultural structures (the bulk of the funds under this Objective) were in rank order: France; Spain; Germany; Portugal; and Ireland. The main beneficiaries for the measures concerned with marketing and processing of agricultural products were Spain, Greece, Italy and France.

A total of 73 operational programmes had been approved under the earlier phase of Objective 5b. The financial envelope for Community assistance to the development of rural areas under this programme was ECU 2607 million for the period 1989–93 (at 1989 prices). The measures under Objective 5b were available to 50 areas and covered 5 per cent of the Community population and 17 per cent of its territory. The average rate of execution for the operational programmes for the period 1989–93 was 95.7 per cent. In general, the implementation of the programmes was regarded as satisfactory (Commission, 1994a).

The assistance under Objective 5b was concentrated on five main priorities;

- Diversification of the primary sector
- Development of the non-agricultural sector
- Development of tourism
- Conservation and development of the natural environment
- Development of human resources.

Diversification of the primary sector proceeded well in France, the UK, Belgium, the Netherlands, and, towards the end of the programme period, Denmark. The development of the non-agricultural sector was successful in Denmark, Belgium, Germany, the UK, France, and, to a lesser extent, Belgium, with the UK and France also succeeding in channelling funds into tourism.

*The new Objective 5b programme*

Under the revised rules for Structural Funds (PIRPAP, Volume Four, pp. 222–3) the following types of areas will now qualify for Objective 5b assistance:

(1) Areas with a low level of socio-economic development which answer two

of the following criteria: high proportion of jobs in agriculture; low level of agricultural income, expressed as value added per agricultural labour unit; and a low population density and/or a trend towards severe depopulation.

(2) Other rural areas presenting a low level of economic development and one or more of the following characteristics: peripheral location in relation to the Union's main growth poles; sensitivity to trends in agriculture, especially in the context of CAP reform; small farms and ageing farming population; pressures on the environment and countryside; hilly and mountainous or disadvantaged terrain; and sensitivity to the impact of restructuring of the fishing industry.

In addition to these criteria, the new rules provide that assistance from the Structural Funds should be concentrated in areas suffering the most severe rural development problems.

The selection of eligible areas took place in four stages. Between mid-September and mid-October 1993 the member states submitted their proposals to the Commission. On the basis of the proposals, taking into account the priorities listed above, the Commission drew up a draft list of eligible areas which was finalised on 21 December 1993. On 19 January 1994 the draft list received the approval of the Committee on Agricultural Structures and Rural Development (composed of representatives of member states) and on 26 January 1994 the Commission formally adopted the list of areas eligible under Objective 5b for the period 1994–9.

Apart from Ireland, Greece and Portugal, which are entirely encompassed by Objective 1, all the countries of the Union proposed areas for Objective 5b status. Altogether the proposals covered a population of 39.6 million, equivalent to 11.5 per cent of the total population in the Union (in the earlier period Objective 5b covered 16.6 million inhabitants, or 5.1 per cent of the total population). According to the Commission (1994b), it sought to reconcile the principles of concentration with the desire to increase the number of eligible areas. Thus two main aims guided the selection of areas: (1) maintaining the intensity of aid per inhabitant; and (2) maintaining at the 1989–93 level each state's share of the people in Objective 5b areas.

Given that the funds available for the period 1994–9 total ECU 6667 billion (at 1994 prices) then up to 26.95 million inhabitants could be covered to the same intensity. The final list of eligible areas (given by country in Table 5.1) slightly exceeds this figure, accounting for 28.52 million inhabitants (8.2 per cent of the total population of the Union). As yet it is too early to make any assessments of the measures included under Objective 5b in its second stage. However, Bachtler and Michie (1994) have already pointed out that the overall increase in the areas covered in the Objective contradicts the aim of concentrating resources in the least prosperous regions. They also argue that the widening of regional disparities in the face of trade liberalisation is

Table 5.1    *Population eligible for Objective 5b in the period 1994–9*

| Member State | No. of inhabitants | % of national population |
| --- | --- | --- |
| Belgium | 448 059 | 4.5 |
| Denmark | 360 119 | 7.0 |
| Germany | 7 725 000 | 9.6 |
| Spain | 1 731 271 | 4.4 |
| France | 9 759 427 | 17.3 |
| Italy | 4 827 805 | 8.4 |
| Luxembourg | 29 972 | 7.4 |
| Netherlands | 799 958 | 5.4 |
| United Kingdom | 2 840 997 | 4.9 |
| Total | 28 522 608 | 8.2 |

*Source*: Commission (1994b)

unlikely to be adequately ameliorated by the new structural measures: 'Even after the 1993 budget increases, with the exception of a few regions in southern Europe, the level of EC regional aid can only be considered marginal in relation to the scale of regional problems' (p. 795). Enlargement of the Union will, they also believe, exacerbate the problems, ensuring that before the end of the new programming period (1999) the Commission will be forced into another fundamental review of Structural policy.

*The Committee of the Regions*

Finally, mention should be made of the Committee of the Regions. This is the EU's newest consultative body, set up under the Maastricht Treaty. It held its inaugural meeting in Brussels on 15–16 March 1994. The 189-member Committee represents the regions and local authorities of the twelve member states at EU level and is to be consulted on a mandatory basis in several spheres, including education, public health and economic and social cohesion. It may also deliver opinions on its own initiative.

At its second meeting it called for relevant local and regional authorities to be represented on the committees set up to approve and supervise Cohesion Fund projects and for aid from the Fund to be concentrated more on environmental projects.

# References

Bachtler, J. and Michie, R., 1994, 'Strengthening economic and social cohesion? The revision of the Structural Funds', *Regional Studies*, 28: 789–96

Commission of the European Communities, 1994a, *The implementation of the reform of the Structural Funds: Fourth Annual Report*, The Commission, Brussels

Commission of the European Communities, 1994b, *Assistance from the Structural Funds for the development of rural areas: Objective 5b areas 1994–1999 Directorate-General for Regional Policies*, The Commission, Brussels

## Further reading

Miles, L., 1993, *Scandinavia and EC Enlargement*, European Community Research Unit, University of Hull, is a useful source of further information on the Scandinavian enlargement

# 6 Towards sustainable tourism for Europe's protected areas — policies and practice

## Rosie Simpson

## Introduction

Holidays are now part of the European way of life. Increased leisure time and disposable income coupled with a fashion for travel mean that millions of Europeans take holidays every year. Now that travel is fast and relatively cheap, even the remotest places can be reached. Today's tourists are increasingly adventurous and want holidays based on nature, culture and outdoor activities. What implications do these trends have for Europe's national and nature parks and for other areas that aim to protect the dwindling resource of unspoilt nature and landscapes?

Mass tourism has already spoilt many attractive parts of Europe, including some protected areas. A number of protected areas continue to experience extreme pressure and others face new developments. Tourists and associated facilities can cause erosion and pollution, bring urbanisation, consume scarce local resources and disturb and damage both nature and local communities. But tourism brings benefits too. It supports employment and contributes substantially to the economy of many countries. Many protected areas lie in rural regions where tourism is one of the few economic activities — sometimes the only one. Post-Communist countries increasingly see tourism based on protected areas as an economic saviour that will help to support conservation. Tourism can help to justify funding for conservation and sometimes contributes directly.

But is tourism really a panacea or is it a poisoned chalice? How can the people responsible for protected areas manage tourism to optimise its benefits? How can they prevent the areas that they are protecting, and which are the foundation stone for tourism, being irreparably damaged?

These questions led the Federation of Nature and National Parks of Europe (FNNPE) to set up a Green Tourism Working Group in 1991 which soon changed its name to the 'Sustainable Tourism Working Group' (the FNNPE

*Progress in Rural Policy and Planning*, Volume Five. Edited by Andrew W. Gilg
© 1995 Editors and contributors. Published 1995 by John Wiley & Sons Ltd.

group). In part this reflected the findings of the 1992 Earth Summit in Rio and the growing consensus that all development needs to be sustainable. It also highlighted concern about the use of terms such as 'green' tourism and 'eco-tourism' which have no agreed definitions. Such terms are sometimes used more as a marketing label than as a genuine indication of environmentally sustainable activity. Thus, in an attempt to overcome these ambiguities the FNNPE group, in their 1993 report *Loving them to death?* (Federation, 1993), defined sustainable tourism as 'all forms of tourism development, management and activity that maintain the environmental, social and economic integrity and well-being of natural, built and cultural resources, forever'.

This chapter provides a three-part commentary on the FNNPE report. The first part describes the context of tourism trends and policies in which the report is set, the second summarises the guidelines and recommendations of the report, and the third part looks at the report's impact in relation to some recent developments in tourism and protected areas.

## Background to the FNNPE report and working methods

The FNNPE report aimed to produce both practical guidance for managers of protected areas and also advice for the tourist industry on ways to make tourism sustainable in and around protected areas. In both cases the guidance and advice were to be based on current good practice. The report did not set out to review European policies and practices by a comprehensive survey, but instead used the experience of the FNNPE professional network as set out below.

The 14 members of the Working Group came from 11 European countries and were experienced in managing tourism and protected areas. In addition, almost 60 professionals took part in three regional workshops to explore issues in: North-West Europe; Southern Europe, Mediterranean and Alpine areas; and Central and Eastern Europe. Forty case studies were also received, from which 16 were selected for the final report.

## Trends in tourism

World tourist numbers grew by 1700 per cent between 1950 and 1991, from around 25 million in 1950 to 450 million by 1991 (World Tourism Organisation, 1992). Leisure patterns in Europe also changed drastically during this period and holidays are no longer considered a luxury. Over 180 million EU citizens take holidays away from home each year. Tourism in Europe employs over 7 million people and accounts for more than 5 per cent of Gross Domestic Product and foreign trade (Commission, 1992a).

Tables 6.1    *Key tourism trends*

(1) A steady growth in tourism in Europe of 3.0–4.5 per cent per year over the next 10 years
(2) Tourism to the Mediterranean region is expected to double in the next 30 years
(3) A significant growth in 'Western' visitors to post-Communist countries
(4) A 45 per cent increase in the number of cars in Europe in the next 20 years (most visitors to protected areas travel by car)
(5) A growing demand for holidays based on nature and outdoor activities and for cultural and educational tourism
(6) An increase in tourism that is 'environmentally friendly'

'Mass' or package tourism grew rapidly in the 1970s and 1980s, particularly in the Alps and the Mediterranean region. However, tourists of the 1990s are demanding a wider variety of holidays, including adventure trips and holidays based on nature and culture. These and other forecast tourism trends suggest that pressure on nature and national parks and other protected landscapes will increase in the decades to come as shown in Table 6.1. World Conservation Union has identified tourism pressures as one of the major threats to protected areas in Europe (International, 1994), and the EU considers that the greatest potential pressure will continue to come from mass tourism in coastal and mountain areas (Commission, 1992a).

There is, however, a dearth of accurate and comparable information about the number of visitors to protected areas at the national and local levels. This makes it impossible either to assess the overall significance of tourism to Europe's protected areas or to identify trends at the national or European scale. Most information relates only to individual areas or countries. For example, the Countryside Commission (1992) has estimated that the 103 million visitor days spent each year in national parks in England and Wales contributed between £550 and £900 million to the local economy, surveys in the Tatra National Park in the Slovak Republic show that visitor numbers have doubled in the last 20 years, and in Exmoor National Park in England although visitor numbers are considered to be stable, activities such as mountain biking and horse riding are increasing and penetrating further into the park. Other assessments of trends rely on professional judgement by protected area managers. These vary markedly. Some are concerned about growing pressures and the need for increased visitor management or greater control, whereas others see potential for further development of suitable forms of tourism, particularly in remoter area and in post-Communist countries. For example, visitors to the Lahemaa National Park in Estonia and the Krkonosce National Park in the Czech Republic are increasing and these areas are under pressure for new major developments of tourist facilities.

## Policies for tourism and conservation

No policies for protected areas or tourism existed at the EU level before 1990. The first policy came in 1990 when a resolution on mass tourism proposed that particularly fragile natural areas needed protection (European Parliament, 1990). The first comprehensive tourism policy was not approved until 1992 (Commission, 1992b). This included a three-year action programme, with tourism and the environment as one of the priorities. The FNNPE project was grant-aided as part of the rural and cultural tourism section of this programme.

In the run-up to the 1992 UN Rio Conference on Environment and Development, the EU published *towards sustainability*, its policy and action programme for the environment and sustainable development (Commission, 1992a). This made no specific reference to protected areas but suggested the creation of buffer zones around sensitive areas. It saw as essential the need to place the future growth of tourism within the framework of sustainability. The elements of a strategy for tourism were set out but little indication was given as to how progress would be monitored.

Over the last 40 years at the national level, most European countries have developed policies and legislation for protecting their most important natural and landscape areas. Countries such as the Netherlands, England, Wales, France and Sweden have relatively strong legislation together with a good network of protected areas. Others are still developing systems, and the 1994 IUCN report, *Parks for life*, suggested that improvements are needed in Albania, Romania, Ireland, Greece and Portugal to ensure that important natural areas become adequately protected through designation (International, 1994).

There are well over 10 000 protected areas in Europe, of which IUCN Category II national parks and Category V protected landscapes account for over three-quarters. In Category II areas, such as the Gran Paradiso National Park in Italy, conservation takes top priority but many of these areas not only attract visitors but also have recreational and educational objectives. Category V areas, such as nature parks in Germany and UK national parks, aim to maintain attractive cultural and natural landscapes and to help people to enjoy them. Many are inhabited and also have a role in relation to social and economic development.

The *Parks for life* report notes that many protected areas are unable to fulfil their objectives because they lack effective policies and legal powers. For example, they lack clear management objectives and need better institutions and resources to achieve objectives. Many protected area managers consulted during the FNNPE project felt particularly helpless in the face of tourism pressures. Tourism was seen as a threat because of insufficient powers to control activities that were damaging and the inability to act quickly to control unsuitable new activities.

In more detail, the FNNPE project found that Italy had no laws controlling tourism, and that the legislation covering Slovenian protected areas was inadequate for guiding tourism development. In contrast, French national parks had strict legal controls that limit tourism. For example, no new facilities can be built, including mechanical lifts, buildings, ski pistes or trails, and the movement of motorised traffic, overflying by hang gliders and the use of mountain bikes were also restricted. Elsewhere in the Czech Republic, public access can be forbidden in threatened areas, and laws covering the Bavarian Forest National Park in Germany enable the park to control visitor access and activities strictly.

By 1992 there were few national policies to guide the development of sustainable tourism related to protected areas and no national strategies for sustainable tourism. There were initiatives in France and Luxembourg but it was in the UK that the greatest progress had been made. In 1991 the government set up a 'Tourism and the Environment Task Force' which established principles for sustainable tourism (English Tourist Board, 1991). Close collaboration between the national agencies responsible for tourism, rural development and national parks led to the agreement of principles for tourism in national parks. This was followed up by a jointly published guide to sustainable tourism in the countryside as a whole (Countryside Commission *et al.*, 1991b), a guide to good practice in the national parks (Countryside Commission *et al.*, 1991b), and principles for tourism in national parks (Countryside Commission *et al.*, 1993). Each national park is covered by a structure plan. Development control powers enable the national park authorities to control many (but not all) developments, including new tourism facilities.

## The need for more strategic action for sustainable tourism

It is essential to have a policy framework for individual protected areas which sets tourism in a regional and national context if the problems outlined above are to be avoided. In protected areas, conservation should be a major objective, if not the primary one, and these areas should not be the prime focus for tourism development.

A key recommendation of the FNNPE report was the need for strategic action including:

- *Stronger legal measures* (both enacted and enforced) to designate and protect nature and national parks and to control tourism development and activities effectively
- *National strategies and policies for sustainable tourism* that set protected areas in a national context;
- *Improved information about tourism related to protected areas and its trends* at the European and national levels.

**Table 6.2**  *Tourism activities generally compatible with protected areas*

Activities that are based on the area's special character and quality
  Appreciating nature
  Cultural and educational tourism
  Quiet, small-scale or small group activities
Activities that cause no damage, disturbance or pollution

*Source*: Federation (1993)

**Table 6.3**  *Tourism activities generally incompatible with protected areas*

Large-scale facilities associated with organised of 'mass' tourism
Activities that are noisy, involve large numbers or that repeatedly disturb wildlife
Skiing and other large-scale sports facilities and events
Motorised recreational activities

*Source*: Federation (1993)

## Developing guidelines for sustainable tourism associated with protected areas

Protected areas vary greatly and one characteristic of sustainable tourism is that it does not come as an 'off-the-shelf' package. It needs to be tailored to the character, needs and capacities of each area, but should follow some general principles. Tourism and associated developments related to protected areas should be in harmony with the local environment in terms of scale, style and character. Accordingly, the FNNPE report argues that sustainable tourism should:

(1) *Be based on the special qualities of protected areas* providing opportunities to enjoy the beauty of their nature and landscapes but not disturbing or destroying the qualities that people come to enjoy, such as peace and quiet, nature and landscape.
(2) *Cause no environmental damage, disturbance or pollution.* Activities and developments and associated elements, such as transport, should minimise pollution, waste production and the use of energy and other resources.
(3) *Benefit the environment, the local community and the tourism sector alike.* Tourism can provide a much-needed boost to the economy, but should not be at the expense of the environment or local communities and the benefits to all sectors should be sustainable in the long term.

Strategic policies should also recognise these principles. Tables 6.2 and 6.3 list the type of tourism activities that are likely to be compatible and incompatible with protected areas.

**Table 6.4**  *Guidelines for sustainable tourism*

| | |
|---|---|
| (1) | State clear aims |
| (2) | Compile an inventory |
| (3) | Work in partnership |
| (4) | Identify the values and image on which to base sustainable tourism |
| (5) | Assess carrying capacities and set standards that must be maintained. |
| (6) | Survey and analyse tourist markets and visitors' needs and expectations |
| (7) | Identify tourism activities that are compatible with the protected area |
| (8) | Propose new 'tourism products' to be developed |
| (9) | Assess the environmental impacts of proposals |
| (10) | Specify visitor management required such as zoning and channelling, interpretation and education |
| (11) | Propose traffic-management systems |
| (12) | Devise a communications and promotional strategy |
| (13) | Establish a programme for monitoring and review |
| (14) | Assess resource and training need |
| (15) | Implement the plan |

## The FNNPE guidelines for sustainable tourism

An action plan for sustainable tourism should be produced for each protected area, using the 15-point guidelines in Table 6.4. The process of producing the plan can be as important as the plan itself. It should enable all the main interests involved locally in tourism — the protected area managers, the local communities and the tourism sector — to identify the most appropriate form of tourism for their area.

In many protected areas, work has already been carried out on a number of these points. The guidelines are designed as a general checklist rather than a step-by-step guide. Some parts need to be implemented in sequence whereas others must be carried out continuously — working in partnership and monitoring, for example. The FNNPE report gives additional guidance for special areas: post-Communist countries; Mediterranean areas; mountains used for winter sports; coasts; and wetlands.

## Examples of the approach

A number of points will serve to illustrate the approach taken by the FNNPE guidelines. These include the need to define clear conservation, social and economic aims. Conservation aims are centred on the fact that tourism in and around protected areas depends on the quality of their landscapes and wildlife. These must conserved, or the future of both the protected areas and of tourism will be jeopardised. Social and economic aims also need to be developed to ensure that benefits to local businesses and communities are optimised.

The successful development of sustainable tourism requires a partnership approach. Local communities, the tourism sector and managers of parks and reserves must work together to create forms of tourism that benefit the parks and reserves and the local area. This approach is well illustrated in the project in the Pirin National Park in Bulgaria supported by the UK Environmental Know How Fund. A local project officer has been appointed to help the local municipalities to work together. He is also stimulating new local initiatives that will contribute to sustainable tourism in the area; benefit the local economy; and provide an alternative to mass tourism based on downhill skiing. This is beginning with the development of bed and breakfast accommodation and small-scale projects such as craft trails.

In the past, the image of protected areas and their tourism potential has been promoted by multi-million-pound marketing campaigns organised by the tourist industry and by media coverage. This can have a massive impact on their audiences and create unrealistic expectations about the facilities that a protected area should provide. Some publicity can give the false impression that rare wildlife and spectacular landscapes can be seen without effort, and wrongly suggests that parks and reserves are places for large-scale hotels and sports facilities. Such images also tend to generate demand for new large-scale facilities or for inappropriate activities. Protected area managers can then only react defensively in the face of the resulting pressure.

In order to shape tourism to the needs and capacity of their areas, managers need to be in a position to take a positive approach. As a first step, they must identify the image and values of the area on which sustainable tourism will be based. These include peace and quiet, undisturbed nature, beautiful landscapes and a healthy environment. Tourism developments and activities should respect and maintain these values. Close collaboration with the media is then needed to promote the values which the parks and reserves aim to protect.

The next step in developing tourism that respects the character of individual protected areas is to generate ideas for new 'tourism products' that are appropriate, and to work in partnership with the tourism sector where possible. These ideas might include educational activities; holidays based on nature and culture; and those reflecting the special local character and qualities of the protected area, such as local crafts, products or cuisine. An assessment should be made of the impact on the immediate and wider environment of any proposals. Those that are developed should minimise their environmental impact and optimise the benefits for the protected area, the local economy and local communities.

There are already examples of successful new developments. In France, the Fédération des Parcs Naturels Régionaux developed a new package of holidays based on national and regional parks, called 'Voyages au Naturel'. The holidays are promoted in a national brochure and include walking, cycling and nature holidays in small groups using comfortable local accommodation and cuisine. To maximise benefits to the local economy, all the services are provided by local tourism operators who must comply with the Fédération's quality charter.

Tourism will always need careful management whenever it takes place in or around protected areas because of their national importance and sensitivity. Some protected areas, or parts of them, are so vulnerable that there should be no tourism development, but usually at least limited access is possible. Effective visitor management techniques must be a key part of any sustainable tourism plan in order to keep tourism pressure below the levels of carrying capacity, and thus prevent damage. In managing visitors it is important to give a positive message and not to appear unnecessarily restrictive.

Education and interpretation programmes have an important role, not just in improving visitor awareness but also in making visitors feel welcome and helping them to enjoy themselves. Other management techniques include: zoning areas and activities or restricting access at certain times; channelling visitors by means of paths and trails or providing 'honeypot' attractions in more robust parts; providing ranger services; and developing special transport systems. There are numerous examples of successful visitor management measures, some of which are included as case studies in the FNNPE report. For example, in the Peak National Park in England car parks have been made smaller to reduce the number of visitors, and footpaths have been carefully reinforced to repair damage and prevent further erosion, and in Austria, where too many cars were using the scenic route along the Grossglockner Pass in Austria's Hohe Tauern National Park, a scheme has been implemented which involves traffic restrictions, differential pricing and the development of bus services to reduce the use of cars and to increase public transport use. The aim is for 60 per cent of visitors to arrive by public transport by the year 2000.

## Wider action needed

In recommending action for protected areas and in giving examples of good practice, the FNNPE report works on the principle of 'think globally, act locally'. However, action in protected areas alone will not make tourism sustainable. Tourism in and around parks and reserves is influenced by broader tourism patterns and itself affects a wider environment. Implementing the guidelines will require political commitment and practical backing, including resources, from international organisations, national governments and the tourism sector.

The improvements needed in international and national legislation and policies have already been discussed. The FNNPE report recommends the following additional action:

- *A European Charter for Sustainable Tourism Operation* setting standards for tourism in and around protected areas, developed in association with the tourism sector and adopted by them.
- *A European Action Programme for sustainable tourism in and around nature and national parks* offering incentives to develop sustainable tourism and to transform existing but non-sustainable development into sustainable forms.
- *Better training opportunities* for the tourism sector, including tour guides in protected areas, park managers and tourism operators, particularly those within local communities.

## Follow-up to the FNNPE report

The *Loving them to death?* report has received widespread acceptance from members of FNNPE and from other organisations. The most significant development at the European level has been endorsement by IUCN of the recommendations for strategic action and of the need for all Europe's protected areas to have management plans that include plans for sustainable tourism (International, 1994). A proposal from the Fédération des Parcs Naturels Régionaux de France is seeking EC funding to develop the European Charter for Sustainable Tourism in cooperation with the tourism sector, and to test its implementation in ten pilot areas. The project would also provide an advisory service to give practical advice on implementing sustainable tourism in and around protected areas.

Protected area managers are already beginning to put the guidance into practice. The Wye Valley Area of Outstanding Beauty, on the border between Wales and England, developed a sustainable tourism plan, involving the tourism sector and other local interest groups in 1994. A principal aim was to test the practicality of the FNNPE guidelines. A plan for sustainable tourism was also being developed in 1994 by the German–Belgian Nature Park, coordinated by the Chairman of the FNNPE Working group. Although not designed specifically as a follow-up project, it used the FNNPE approach.

Others are keen to learn more about how to develop an action plan, before they begin work. Thirty protected area managers from 21 countries attended a training course in September 1994, held at the Peak National Park Centre. This should stimulate practical projects across Europe. A second course was planned for 1995.

## Room for improvement in recent policy developments and initiatives

There is still a need for clear and detailed policies for sustainable tourism at the EU and national levels that set tourism in protected areas in a broader context. Within the EU, there are unlikely to be new EU policies addressing the effects of tourism on the environment in the near future. Better use will be made of existing measures such as environmental directives and structural funds. The Commission was due to issue a Green Paper at the end of 1994 (European Commission, 1994).

Detailed sustainable tourism policies have yet to be developed in most countries, although an outline should be included as part of strategies for sustainable development in line with the recommendations of the Earth Summit in Rio. The 1994 tourism policy for Wales sets a good example of a more detailed policy. It proposes the vision for a tourism industry that 'is sustainable in the long term in both environmental and economic terms and is integrated within the community' (Wales, 1994). It also includes specific policies for protected areas. At the national park level, the North York Moors National Park Authority has published tourism policies (North, 1991) backed up by detailed guidance for tourism businesses showing the type of tourism likely to be appropriate in the park (North, 1993).

Sustainable tourism is currently a popular theme for international and national funding programmes, and it is a priority action for the 1994 Life Programme (Commission, 1994). The EU also promotes a range of programmes that are being used to fund sustainable tourism projects, sometimes within protected areas. For example, the PHARE programme, to assist the economic restructuring of countries in Central and Eastern Europe, supports an ECU 8 million tourism development project in Poland. However, an assessment of the environmental impact of the PHARE programme as a whole observed that no project had been subject to an environmental impact assessment (European Parliament, 1993). It also highlighted the policy vacuum in which many of the projects operate.

The World Wide Fund for Nature (WWF) supports CADISPA (Conservation and Development in Sparsely Populated Areas) with projects in Italy, Greece, Scotland, Spain and Portugal (World, 1994). International funding agencies, such as the World Bank and the European Bank for Reconstruction and Development, also fund tourism projects related to protected areas. The IUCN recommended that these and other bodies should employ strict measures to ensure that projects they fund do not harm either the environment or protected areas (International, 1994). They also proposed that the EU should re-examine its approach to programmes like PHARE, to ensure that sustainable development is an aim.

Many national initiatives are also funding sustainable tourism projects that affect protected areas, including a sustainable tourism project in the Green Lungs area of Poland funded by WWF and the UK Environmental Know-How Fund (National, 1993).

In general, these examples have shown that current initiatives lack an overall policy context except for the vague exhortation to be sustainable, and in detail, that strict control measures like environmental impact assessment are the exception rather than the rule. There is thus much room for improvement.

## Conclusions

Emerging policies and the many initiatives being developed are a positive sign that sustainable tourism development is much higher on the agenda than it was a few years ago. The FNNPE project nonetheless found that most existing tourism is not sustainable, even in protected areas, although the case studies did show that progress is being made with small-scale projects. Elsewhere however, Jenner and Smith (1992) have reported that only 10 per cent of eco-tourism projects actively involved reducing impacts and waste to a minimum. Positive examples are nonetheless becoming more common. There are now solar-powered mountain huts and gîtes in some of the French regional and national parks. A holiday centre in the Sierra del Norte Natural Park in Spain treats and recycles waste water and uses water power to generate energy.

However, four key improvements are still required. First, there is an urgent need for demonstration projects to show real sustainable tourism in practice. Such sustainable tourism should: respect the character of local areas; contribute to the local economy; not damage the local area or the communities that live there; and minimise effects on the wider environment, such as resource use, pollution and waste production. At present, few tourism projects take all these factors into account. Second, there should be a proper appraisal of the environmental, social and economic effects of all tourism developments and activities associated with protected areas. Third, mechanisms are needed to exchange experience gained from the many current projects that aim to demonstrate sustainable tourism. At present, the findings are uncoordinated and lessons and good practice are not being communicated to others. Finally, more national and regional policies are needed to make sure that only appropriate tourism developments take place in protected areas and to help focus funding.

Tourism based on nature and national parks is likely to increase in the years to come. Properly managed, it should bring benefits for conservation, the local economy and local communities. However, tourism is not a panacea and should not be accepted at any price. Some characteristics of the tourist industry may be difficult to reconcile with managing protected areas. For example, the dominance of the private sector, which tends to aim for maximum profits, can make it difficult to limit capacity and visitor numbers (McKercher, 1993). In the absence of a strong and diverse rural economy, there is also the danger that some protected areas will become too dependent on tourism, leaving them vulnerable to any change in the current fashion for holidays in protected areas.

Tourism in and around the parks depends on the high quality of their environment. If tourism is to successfully co-exist with the conservation of protected areas, it must be sustainable in the long term. This applies both to new developments and activities and to existing non-sustainable tourism that needs to be transformed into sustainable forms. In areas that are already under severe pressure, tourism impact may need to be reduced. These are difficult tasks but ones which people are beginning to address. Success will increase people's enjoyment of the parks, create stronger local communities and economies, and, above all, safeguard Europe's most precious wildlife and landscapes, which are the foundation stones for tourism in these regions.

*Author's note*

Rosie Simpson works for the National Parks and Planning Branch at the Countryside Commission in England and was seconded in 1992–3 to the Federation of Nature and National Parks of Europe, to coordinate the Sustainable Tourism Project. Please note that the information in this chapter is based on the FNNPE report and does not reflect the policies or views of the Countryside Commission.

# References

Commission of the European Communities, 1992a, *Towards sustainability — EC programme of policy and action in relation to the environment and sustainable development*, The Commission, Brussels

Commission of the European Communities, 1992b, 'Council decision of 13 July 1992 on a Community action plan to assist tourism, 92/421/EEC', *Official Journal of the European Communities*, No. L 231/26

Commission of the European Communities, 1994, *Life 1994 information package*, The Commission, Brussels

Countryside Commission, 1992, *Fit for the future — report of the National Parks Review Panel*, CCP 354, The Commission, Cheltenham

Countryside Commission, Countryside Council for Wales, English and Wales Tourist Boards and Rural Development Commission, 1991a, *Tourism in national parks — a guide to good practice*, The Commission, Cheltenham

Countryside Commission, English Tourist Board, Rural Development Commission, 1991b, *The green light — a guide to sustainable tourism*, The Commission, Cheltenham

Countryside Commission, English Tourist Board, Rural Development Commission, 1993, *Principles for tourism in national parks*, The Commission, Cheltenham

English Tourist Board, 1991, *Tourism and the Environment, Maintaining the balance*, The Board, London

European Commission, 1994, *The week in Europe*, 19

European Parliament, 1990, *Report on the measures needed to protect the environment from potential damage caused by mass tourism*, Document A 3–0120/90, The Parliament, Brussels

European Parliament, 1993, *An assessment of the environmental impact of the PHARE Programme*, Project paper No. 1, The Parliament, Brussels

Federation of Nature and National Parks of Europe, 1993, *Loving them to death? Sustainable tourism in Europe's Nature and National Parks*, The Federation, Grafenau, Germany

International Union for Nature Conservation, 1994, *Parks for Life — action for protected areas in Europe*, The Union, Geneva

Jenner, P. and Smith, C., 1992, *The Tourism Industry and the Environment*, Economist Intelligence Unit, Special Report, No. 2435, The Unit, London

McKercher, B., 1993, 'Some fundamental truths about tourism: understanding tourism's social and environmental impacts', *Journal of Sustainable Tourism*, 1: 6–16

National Foundation for Environmental Protection, 1993, *Green Lungs of Poland*, The Foundation, Bialystock, Poland

North York Moors National Park, 1991, *National Park Plan, Second Review*, The National Park, Helmsley, England

North York Moors National Park, 1993, *Landscape for visitors – sustainable tourism in the North York Moors National Park*, The National Park, Helmsley

Wales Tourist Board, 1994, *Tourism 2000 – a strategy for Wales*, The Board, Cardiff

World Tourism Organisation, 1992, *Tourism Trends and the Year 2000 and Beyond*, Presentation for the World Trade Centre, Seville

World Wide Fund for Nature, 1993, *CADISPA Project Portfolio*, The Fund, Brussels

# 7 The changing competitive advantage of rural space

## *Elena Saraceno*

## Introduction

In this chapter some of the implications of the Italian industrialization process are examined and are shown to illustrate some general features of development in rural areas. It is argued that the recent success of local economies is due to new trends in global markets, as well as Italy's historically rooted territorial differences, which have heavily influenced the way in which industrialization has taken place. Because of its link with the current processes of restructuring of the world economy, the Italian case cannot be considered exceptional, but rather as a sort of modern laboratory in which most 'paths' towards industrialization have been tried. We can thus discern the two major and interconnected trends that characterize spatial differentiation today as the shifting location of economic activities towards new regions and the increasing social and economic diversification of rural areas. New options for rural development are being opened up, based on economies of diversification, changing comparative advantages over time, and the growing segmentation of global demand. However, the ability of rural areas to respond successfully to these new trends varies considerably, depending on the legacies of past forms of development. Four types of rural scenario are described below, and the different responses that each may give in order to increase their competitiveness are assessed. In the final section, the view that rural areas can no longer be seen as necessarily declining is proposed. If this is true then future tasks include a rethinking of conceptual tools and indicators of rurality and a major reformulation of current rural policies.

## The case of Italy: an exception or a laboratory of recent changes?

Italy has followed in its industrialization process a peculiar and viable pattern, both in its regional location (space) and in its stages of development

*Progress in Rural Policy and Planning*, Volume Five. Edited by Andrew W. Gilg
© 1995 Editors and contributors. Published 1995 by John Wiley & Sons Ltd.

(time). Up to the 1960s these peculiarities were interpreted mainly on the basis of the development–underdevelopment perspective, applied to North–South differences. The 'most advanced' area was the North, while the South lagged behind. The implication was that with time and appropriate policies (there were different schools of thought about this) the South would catch up with the North. The path towards development for the lagging area was a sort of imitation of the stages through which the advanced area had already passed. It was assumed that there was only one path towards industrialization. This perspective *de facto* eliminated space and time as crucial variables of development: once a typical sequence of stages was established, generalized from the experience of the most developed countries, there was no need for more information. There were only two types of situation, the developed and the underdeveloped, with few significant differences within each category. This homogeneity in turn made possible relatively standardized territorial development policies.

In the late 1970s a new consideration revealed the possibility of more than one path towards industrialization and initiated a debate over whether patterns of development were convergent or not. This new consideration was the empirical discovery of what was then called the third Italy (the Centre and North-eastern regions); an area where a process of industrialization had been achieved without major agrarian change. The so-called third route to development was based on the maintenance of the pre-industrial artisan enterprises (more often with light rather than heavy industry); it was achieved without significant amounts of capital (already existing technology was adapted to new uses); and included the invention of a new form of organization of production (based on the cooperation and exchange of small and medium enterprises, together showing economies of scale similar to those of large enterprises, greater flexibility in the presence of economic downturns and competitive advantages on the global markets). The implication of this third way for the lagging South was that there could be more options for development than previously allowed. Relevant policies to promote one or the other path could vary substantially.

Furthermore, the possibility of multiple paths towards industrialization reintroduced the need to consider space and time. It was, in fact, possible that some areas, given their pre-industrial characteristics, could follow more easily one type of development trajectory rather than another. For example, if, in the South, the persistence of large farms, the lack of a local artisan structure and the different institutional and organizational structures had indeed deterred the endogenous development typical of the third Italy, more attention should be paid to the specific differences within underdeveloped situations, which meant distinguishing among types of backwardness. In this way, spatial differences became again meaningful dimensions of development policies.

If specific pre-industrial conditions are believed to be an important explanatory factor in accounting for the observed differences in the patterns of industrialization, we are in fact attributing importance to *history*. But we might also reflect on the contemporary reasons for the appearance of this new pattern of development. In other words, why have industrial districts, which Alfred Marshall described for England in the early stages of industrialization, reappeared as a competitive organization of production in Italy today? The answer to this question lies not only in the pre-industrial conditions but also in the existence of the global economy. It is the combination of these which explains the competitive advantage of small and medium industries today.

Italian success in late economic development has been attributed both to the peculiarities of historical conditions and to the increasing segmentation of markets. Many observers have dismissed the problem posed by the presence of two patterns of development at different times (in the first half of the century the first pattern and in the 1970s and 1980s the second) and in different areas of the country (one in the North-west, the other in the North-east and Centre), by attributing them to the differences of the pre-industrial economic structure rather than to current trends in the economy. In this way, the peculiarities of Italian development are perceived as exceptions which do not question in any way the general validity of dualistic thinking. There might be more ways to industrialize than expected but the end product is believed to be the same, namely that, with time, small enterprises increase in size and concentrate, thus becoming classical industrial organizations just as happened for the early comers. From this perspective the Italian peculiarities are just the exception that confirms the (dualistic) rule.

On the other hand, if we imagine that the globalization of the economy has been accompanied by an increasingly segmented and changing demand for products, the recent success of small and medium enterprises might be better explained as the modern response to new market conditions. For these conditions require short production runs to which large enterprises, with an organization and technologies developed in a period of mass consumption, may have to adapt. From this perspective the third Italy should not be interpreted as an exception which, in due course, will evolve towards concentration but, on the contrary, should be seen as the vanguard case, facilitated by particular historical conditions, that have furnished it with the ideal resources and capabilities to be the first to exploit aspects of the current processes of economic restructuring.

In the 1980s and early 1990s, these two schools of thought have been looking for arguments to support their respective interpretations, and the evolution of the economy seemed to provide justification for both positions. At first, the evaluation of the role of small and medium enterprises in the contemporary restructuring of the economy was left to Italians (Bagnasco, 1978, 1988; Becattini, 1987; Brusco, 1989), but later this debate spread to American and other European industrial economists and has merged with the analysis of the crisis of Fordism (Piore and Sabel, 1987; Boyer, 1986, 1988; Dore, 1988; Pyke *et al.*, 1990; Porter, 1991; Salais and Storper, 1993). In this way the discussion has rightly been set in its proper context — the industrialization process and the changing organization of production — away from strictly national readings. The issues raised are far from resolved, but for the purposes of this chapter it is enough to point out some of the most interesting findings for rural analysis:

- Industrialization is not a definitive state and there is no such thing as a self-sustained economy. Deindustrialization may occur and transform drastically the importance of some centres in relation to their peripheries. Spatial (or regional) shifts in the location of economic activities can redefine the advantages and disadvantages of various territories and cause the rise of newly developed areas, regions and countries and the decline of others. This process might include also the rearrangement of traditionally established relationships between rural and urban areas, however defined.
- The central concept that allows the comparison and evaluation of different industrial systems is their competitiveness in relation to any segment of the global market (which, of course, includes local and regional ones). This appears just as relevant (perhaps even more so) than the adoption of a specific technology, size of organization or sector of activity.
- Spatial differences are far from irrelevant even with the globalization of markets: each area or region is able to compete better if it is able to integrate local, historically accumulated knowledge and capacities with information and exchange about non-local markets and technologies. 'Bottom-up' or endogenous development, if integrated in this way, appears to provide competitive advantages, and for this reason the globalization of markets is more likely to increase territorial differentiation rather than its homogenization.
- Much more attention than in the past should be paid to the evolution of consumption patterns within industrial societies. Mass consumption is in fact turning out to be only one phase and not necessarily the final one in the industrialization process. The segmentation of global markets, already mentioned, seems to refer to increasingly differentiated consumption patterns which reduce the market size of many goods and their durability. This creates a need for a more flexible production organization for which medium and small enterprises, linked among themselves in more or less formal networks, achieve efficiency more easily then large ones.

The multiple ways in which Italian areas have experienced economic development should be considered not an exception but an indicator of how different areas might combine in specific ways their peculiarities with present market demands, thus differentiating rather than imitating the path followed by earlier industrialized areas. Furthermore, precisely because late industrial development maintained consistent human resources in rural areas, the most recent industrialization process has shown a diffused pattern over the same rural areas. On the one hand, this has reduced the classical division of labour between town and country, and, on the other, has revitalized the extremely rich structure of cities and towns of different sizes to be found in Italy, and which date back to the Middle Ages (Saraceno, 1994).

Such recent development could be considered an example of the regional shift of activities between different areas (the rise of the North-eastern regions) and within them a redefinition of the urban and rural roles which was brought about by diffused industrialization. These regional shifts do not correspond to any hierarchical organization of space in relation to previously developed regions or centres, in a spillover pattern, and do not reproduce the expected dynamics between development and underdevelopment (dependency of the emerging area). On the contrary, they represent a new spatial organization whose relationship with old centres cannot be predetermined and must be specified in each case.

## Shifting regions and the diversification of rural areas

If the Italian case can be seen as an important laboratory of changing trends in the global economy, in relation to different geographical areas and forms of production organization within national boundaries, the schematic analysis presented in the previous section allows us to explore not only regional shifts in the location of economic activities in other countries but also the process of economic diversification in rural areas (Bagnasco, 1994). The processes outlined in the previous section which are responsible for the increasing competitiveness of rural areas are the most important of all the processes currently under way.

Observers have pointed out the presence of a marked, long-term trend, consisting of a movement of industrial activities out from older industrial areas towards either different regions in the same country or to other less developed countries. This geographical redistribution has not been interpreted in general as economic decline, partly because the service sector and different types of industries have been gradually replacing the activities that moved out and partly because the welfare system has acted as a 'shock absorber' for the unemployed. I have found evidence that France, England, Denmark, Germany and Austria have measured this kind of mobility within their national boundaries during the 1980s and early 1990s. It appears as part of the same process at another spatial level in the emergence of newly industrializing countries (NICs) in the Far East and in the USA former leading industries, such as steel, moved first to other states and then to Third World countries.

The most frequent explanation for these shifts has been that mature industries find lower labour costs in less developed countries. We could also add that emerging countries are the only markets for certain goods with low quality and prices; that some technologies are very easily sold to aspiring local entrepreneurs by developed countries; and that it is certainly preferable, from this point of view, to export capital than to sustain the social and economic costs of importing labour. But a picture of emerging regions based on their poor, second-best characteristics would be a highly distorted one. Higher segments of certain markets are solidly held by some of them (the example of the NICs is illuminating) and they have achieved competitive advantages in many goods destined to the advanced countries. The multiplication of producers in a growing number of countries reduces the size of the potential market that any one enterprise might have, and this cuts the ground from under large multinational firms. The list of factors affecting the regional shift of industrial activities is certainly far from complete, but there is no doubt that this type of spatial mobility has not run into any of the barriers that were imagined in the 1960s and 1970s by theorists of development. As a result, the achievement of competitive advantages on the global market is no longer a privilege of the older industrialized countries, and the nature of the links between areas is an extremely complex matter that should be studied without hierarchical prejudices.

The spatial impact of regional shifts in the location of economic activities has affected urban centres more severely than rural areas in developed societies. This is due to the simple fact that population and non-agricultural jobs were concentrated in large urban centres and metropolitan areas early in the industrialization of many countries. Later regional shifts often favoured development in rural areas, and thus rural areas have been less badly affected by changes in the pattern of industrial employment, except where economic activities have moved first to rural areas, only to then move to other countries, in a rural or urban location.

The shifting location of economic activities is intimately connected with the changing nature of rural areas. Before exploring this matter, a point of clarification is useful. Most of what has been said in the past about the relationships between developed and less developed countries is often found mechanically repeated and applied to urban and rural areas. The difference between the two pairs of concepts is, however, quite substantial. The rural–urban polarization distinguishes homogeneous categories. All rural areas are assumed to have similar ideal characteristics: low density of population; reliance on agricultural activities; traditional value systems; and little differentiation in social organization. At the other end of the scale, urban areas base their homogeneity on a complementary set of characteristics: a high concentration of population; a prevalence of industry and services; the presence of modern value systems and a highly articulated social structure; the site of institutionalized power and bureaucratic organizations; and the setting for cultural contacts and exchange. The developed–underdeveloped dichotomy does not refer to a homogeneous set of criteria but to a functional area (or region) as a whole, comprising its rural and urban parts. The criterion of spatial differentiation is in one case the degree of economic development while in the other it is the physical concentration of people. Movement between the two extremes of the polarization depends in each dichotomy on a specific dynamic. If this is true then why are the two confused? The answer lies in the assumption that pre-industrial societies are agrarian, and because rural areas in developed contexts are left, after a process of urban industrialization, with a declining population. Both underdeveloped areas and rural areas were always thought to need policies in order to achieve general economic development in the first case and to maintain a comparable income in the second (i.e. modern farms are considered to need protection in order to continue producing). These are the main reasons for the confusion between these two quite different concepts and territorial differentiations.

It is necessary to clarify the meaning of these concepts because the shifting location of economic activities and the process of diversification of rural areas redefines the advantages and disadvantages of old and emerging regions as well as the division of labour between rural and urban areas. Emerging areas do not seem to reproduce the classical pattern of urban industrialization but, as was seen in the Italian case, use the space available more freely, following more closely the spatial desires of the entrepreneurs involved, rather than any other more sophisticated location criterion. This seems irrational if looked at through the lens of location theory. However, it becomes quite reasonable in the light of what has been said about the local economy's competitive advantage, acquired by entrepreneurs through an original integration of local capacities and resources (contextual knowledge), which is then integrated with information about the relationships with global markets and technologies (formal codified knowledge) (Becattini and Rullani, 1993). The process of diversification of activities in rural areas is therefore the result of a quite flexible and not necessarily urban interpretation of the best location for non-agricultural activities (industry and services).

We can identify two distinct periods in the diversification process. In the post-war period rural areas in advanced economies lost significant numbers of farmers. However, the decline in the population of rural areas has been less severe than the loss of agricultural labour might imply. This indicates that although some farmers left agriculture, many were able to find a job in other sectors, thus maintaining their residence in rural areas. A gradual and modest diversification process can be measured by the decreasing percentages of agricultural employment in relation to total employment in rural areas. While in the 1950s a rural area might have been characterized by over half of the active population employed in agriculture, today we would be lucky to find a rural area with a quarter of its population employed in this sector. Since this is a relative measure, it also indicates the growing incidence of non-farming activities in rural areas. This long and gradual process of spontaneous diversification of the activities of the local resident population was largely ignored both as a real process and as an interesting policy orientation, against all evidence, until the European Commission recognized the trend and its usefulness for rural development in the second half of the 1980s (Commission, 1988). This long period of blindness corresponded to the political relevance of the rural-agriculture myth and to the strong assumption that farmers had no income alternatives in rural areas. In addition, any recognition of such diversification would have implied a radical reassessment of the compensatory approach.

This first type of endogenous diversification — modest, long term and invisible and often taking the form of pluriactivity — was followed, from the 1970s onwards, by an acceleration of the process. Of course, this general statement simplifies significant territorial differences in the intensity of the acceleration. The shift in the regional location of activities, in the same period, brought about new economic opportunities in rural areas. The diseconomies of agglomeration in urban areas increased the demand for rural housing and for transportation facilities to urban labour markets. The growing sensibility towards environmental protection and sustainable forms of development added a new dimension to more traditional rural–urban exchanges. The segmentation of demand included tourist and leisure activities in the countryside. These changes created a new demand for rural space from the non-rural population which created new opportunities. In some cases these benefited the original local population, in others only the newcomers and non-local enterprises. In any case, this external and visible demand for space, and for living, working and transportation facilities, brought about a social restructuring of rural communities, sometimes perceived as a welcome integration, sometimes bringing quite opposing views of the world, giving rise to social conflict.

The acceleration of the social and economic diversification of rural areas complicates their homogeneous rural character in relation to urban areas, and questions the relevance and significance of the rural–urban dichotomy. The more the rural is diversified and integrated not only with the urban but with the global economy, the less rural in ideal terms it becomes. A new label would satisfy conceptual clarity but runs against the need to retain the rural ideology, which is often at the origin of new demands for rural space. It should be made clear, in any case, that for the purpose of analysis and policy formulation the incidence, dynamics and type of diversification are crucial variables to take into consideration. It is possible that the process of diversification will turn out to be the real modernization of rural areas, interacting in still-unexplored ways with the recent modernization of the agricultural sector.

Past processes of development will inevitably influence the response of rural areas to regional shifts and diversification. Little research has been done with this perspective in mind. In the next section the identification of alternative possible situations will be elaborated upon. For the future it should be clear that the two processes just described are currently operating to change the normal logic of spatial differentiation in a significant fashion.

## The growing competitiveness of rural areas

The combined effect of the two processes described above has produced a relative improvement in the competitive advantage of rural areas. However, a point of clarification is perhaps appropriate concerning the conceptual framework for understanding the real processes that have actually taken place. What has been difficult to explain within a threefold framework containing economies of scale, the assumption of unchanging comparative advantages, and a uniform, mass demand for goods, becomes possible — at least in principle — when we consider a different conceptual framework, based on what have been called 'economies of scope' or diversification (Chandler, 1994). This framework postulates changing comparative advantages over time and a growing segmentation of demand. With these new conceptual tools the growing competitiveness of rural space can get a first tentative, general explanation. In order of time, the presence of alternative logics of production appeared first with the empirical acknowledgement of the success and modernity of small and medium industries in a global market. Such success was supposed to belong mostly to large enterprises, but eventually — in the face of intransigently 'disobedient facts' (Becattini and Bianchi, 1987) — formerly dominant theories were questioned and reformulated. The new conceptual tools provide the framework for understanding those cases which, far from being exceptional, are instead concrete examples of the existence of new forms of competitiveness. Having established the new processes under way, on the one hand, and a conceptual framework, on the other, we may proceed to assess the more favourable — or neutral — impact that these processes may have on rural areas in relation to previous processes, which favoured urban locations. In other words, rural areas now have a chance of development without necessarily following the pattern established by urban areas in the early industrialization period.

We should not make the mistake of thinking that because we have found cases of spontaneous rural development and new tools to explain them we may expect rural development everywhere. Specific historically accumulated conditions and human capacities, as well as the impact of economies of scale, will continue, as always, to pose formidable constraints on the opportunities for rural change. But what is new is that the evolution of the industrialization process has determined a reorganization of production, still under way, which does not discriminate against the location of economic activities in rural areas as previously. Rural development becomes a real option in this new scenario and it may also be promoted using different policy instruments from those of classical development. In this way rural areas can compete autonomously — if they still have the resources or if they can attract them — with urban areas, just as industrial districts can compete with large firms. There is no justification for hypothesizing *a priori* a hierarchical subordination of one to the other. By the same logic that says they can compete, they can also cooperate as a local integrated rural–urban economy in the first case and as networks of enterprises in the second.

Within the logic of economies of scale rural areas were conceived as born losers, while now they can become winners (as they have in fact done in some cases) adopting the logic of economies of diversification. Such economies are particularly well suited to rural areas. The competitiveness in this case is sought with the production and distribution of different products by one or more enterprises instead of seeking it on the basis of one product by a large enterprise.

In past years an intense debate about the recognition and evaluation of the increasing competitive advantage of rural areas has taken place, often touching highly emotional chords. Some have started talking about a renaissance, a '*nouveau ruruax*', diffused industrialization, and counter-urbanization, while others have recognized the change but have seen it as limited to exceptional cases (the Italian experience), as a temporary passage (the United States), or as the result of decentralization processes: looking for cheap labour; involving the black economy; bad working conditions; and, on the whole, unable to modify significantly the overall declining trend of the rural world. The first group has been labelled as optimistic, the other as fatalistic or conservative. The possibility that rural areas might have a chance of development not so much from traditional rural policies but from quite spontaneous processes has been quite difficult to accept and has caught many observers and policy makers unprepared. In particular, increases in competitiveness based on the diversification of rural activities rather than on agricultural specialization has been quite painful to acknowledge for those who spent the post-war period trying to create a rural world based on economies of scale in the agricultural sector, in spite of obvious limitations in the European context. Furthermore, the failed attempt to achieve rural prosperity by agricultural specialization proceeded with the support of increasingly costly policies.

The debate should now proceed less emotionally. First, from a global perspective what we consider an optimistic position may in fact be much less so. Beyond the relative increase in the competitive chances of rural areas, the process of overall economic restructuring might end up with the old advanced societies having a smaller share of the global market, with all the negative consequences that this possibility will have on employment, average income, and the capacity to generate surpluses for redistribution through fiscal, welfare and regional policies. This may happen if the regional shifts in the location of economic activities transcend national or European boundaries and create new and more successful centres of competition elsewhere. Second, the growing competitiveness of rural areas may be offset by the achievements of cities in sizing down. Third, the social restructuring taking place in rural areas as a consequence of diversification may be expected to produce more social conflict and a significant transformation of their *Gemeinschaft* character. Fourth, diversified rural areas have yet to come to terms with the difference between preserving the environment from agricultural activities, rather than from those non-agricultural activities no longer confined to urban areas. There is little room for optimism here.

It has been stated that past processes of development (early or late, concentrated or diffused) will inevitably influence the response of rural areas to regional shifts in the location of economic activities and diversification. Even though little research has been done on this subject it is possible, in the light of the preceding analysis, to try to identify the situations that are likely to emerge. Very schematically, and oversimplifying for the sake of clarity, there are four typical situations or rural scenarios which may arise from the increasing competitiveness of rural areas. Both the regional context and the role of rural areas are considered.

*1. Early industrialized regions where rural areas have been successfully modernized through the transformation of the agricultural sector*

The rural–urban division of labour as a result corresponds to an almost perfect sectoral differentiation. The rural areas of these regions have the lowest density of population since, for a very long period of time, they have been losing those residents that could not find employment in farming activities. There are scarce human resources and low pluriactivity among farmers. Potential diversifiers could only come from the existing farmers' ranks and their families. There has been a low probability of this happening since in such areas, precisely because of the successful modernization of agriculture, there are large farms which from experience are the least interested in diversification. On the other hand, large urban areas concentrate the majority of population. Rural social renewal may come from the attraction of residents from urban centres (rural areas as so-called commuter zones). From this social group there might be more interest in taking up the opportunities offered by diversification, since this would eliminate commuting and may satisfy entrepreneurial desires. It is also becoming apparent that this group, which can stimulate diversification using the environment as a resource, is particularly concerned with environmental preservation. If the latter concern becomes dominant a rigid conservationism may be imposed which effectively prevents all diversifying initiatives. Thus it is in these areas that the greatest difficulties may arise in responding to the most recent opportunities. Moreover, if these type of areas are beyond commuting distance of any city, isolation may result in social desertification.

### 2. Early industrialized regions where rural areas have diversified spontaneously over time, in one or more sectors of activity

In this case the agricultural sector has modernized, but has maintained farms of different size. Pluriactivity is frequent. Human local resources are varied and rich, there is a higher density of population and, in general, there has been demographic stability or slight gains in population due to positive net migration, often offset by negative natural balances. Potential diversifiers are already part of the local population but may also be found among commuters. The long experience with diversification has produced exchanges of labour and sometimes of entrepreneurial capacities between different sectors of activity within the rural area. There is a low sensitivity to environmental concerns. A crucial factor for assessing the potential response to opportunities for continued diversified growth in the area is to adapt its competitiveness to the new rules of an increasingly segmented demand. The presence of large, old factories or the dependence upon external headquarters of local enterprises (the so-called branch plant phenomenon) indicate less local autonomy and medium/high difficulties in adjusting to more recent forms of competitiveness. If this readjustment fails the situation is comparable to that of declining cities.

### 3. Non-developed regions with diversified rural areas

Non-farming activities have in this case a pre-industrial organization based on small-scale, domestic artisan crafts, production for the local market, and traditional exchanges with an urban centre. There is a polarized farm structure and diffused subsistence farming. Pluriactivity is extremely diffused but non-farming activities are usually precarious and seasonal. Demographic pressure on land is very high and so is population density. Outmigration of excess population is combined with positive natural balances. All sectors show low levels of modernization. Even though there are potential human resources for diversification, there is little knowledge of global markets, current technologies, alternative methods in the organization of production, and a low sensitivity to environmental concerns. Here the problem is to find a way of integrating local contextual knowledge with the codified language of industrial society and global markets. Returning migrants have been in the past, and could be in the future, good endogenous entrepreneurs, providing the necessary link between the local and the global. Possibilities for competitive growth are good in the medium of long term.

## 4. Late development, diversified regions

These industrial areas are based on systems of small and medium enterprises, linked into global markets, often specialized in the same sector of activity, and both competing and cooperating in the process of production. The local economy is relatively autonomous from large urban centres although geared to non-local markets. Population density is medium/high, there is a positive migration balance (small flows), and stable or slightly declining natural growth. The agricultural sector has modernized keeping a differentiated structure and pluriactivity and contracting out work are widely diffused. The process of modernization has successfully integrated local contextual knowledge with global codified languages. We find technological innovation, a high rate of creation of new enterprises, and low sensitivity to environmental concerns. Diversification and specialization in some products have developed spontaneously. In these areas the competitive advantage of rural areas is high. The main problem is how to maintain them with rising costs.

In conclusion, this preliminary exercise in identifying areas which might react differently according to their past experience of development and socio-economic organization has served the purpose of showing the kind of reasoning that is possible, once we have identified both the nature of the processes under way and the conceptual tools that could help us to understand better change in rural areas. A recognition that the alternative options outlined above have created new forms of competitiveness should also be an important basis for further discussion about the future of rural areas, without becoming bogged down in spurious disputes, such as that between optimists and pessimists over the chances of rural areas, as if this was essentially a matter of opinion.

## Some conclusions: rearranging spatial hierarchies, indicators and policy measures

The implications for our conceptual and policy approaches to rural areas of the above can be summarized in the following way.

First, we should definitely abandon the usual crying game about the terrible fate of rural areas, for many of them are alive and well. A lot more can be learned from an unprejudiced analysis of how they have achieved their current levels of development and how they manage to maintain their competitiveness. This could help the scientific community to fashion new conceptual tools.

Second, once it is acknowledged that rural areas may be competitive within present trends in the economy, a major task lies before us in terms of rethinking traditional spatial hierarchies and how we might choose good rural indicators. Present spatial differentiation is embedded in functional or structural categories of thought, characterized by hierarchical relationships such as core–periphery, developed–underdeveloped, marginality–centrality, territorial disequilibria, dependency, and disadvantages. These are all concepts which transform the process of change into a mechanical exercise where time and space have disappeared as relevant dimensions. The important thing is not to predetermine the path and the outcome of the socio-economic transformation of rural areas, and to admit the reversibility of acquired advantages as a normal part of what has to be explained.

Current indicators of rural development should abandon the assumption of measuring rurality through agricultural activities, poverty and distance. Instead, as has been proposed in recent work, rural–urban differentiation should always be analysed within its regional context (Organization, 1994). In fact, rural indicators should not be imagined as an urban opposition, but rather – and more positively – by what is considered distinctively rural, which might refer to more than one type of area. From what has been said above, information about the degree and type of diversification, both of activities and population, should be central. Levels of integration with the outside world should be more important than distance. After all, physical mobility is not the only type of exchange with non-local markets. New forms of communication and exchange should be included. Moreover, entrepreneurship, type of sectors present, forms of organization and cooperation among enterprises, participation in networks, and market access are likewise important components of success. Population density, while significant, should be kept as a descriptive rather than a classificatory variable. The threshold between rurality and non-rurality becomes totally arbitrary if measured using density, since it is a function of the type of development (more of less concentrated) and of demographic evolution which have been influencing variations in density much more than rural–urban constitutive characters. We know very little about the non-farming population in rural areas. The percentage of agriculturally active residents could have a different meaning if considered in the light of the possible increase of industrial and service employment associated with absolute variations in the number of farmers. In sum, indicators in general should measure the whole rural world instead of a declining aspect of it.

Third, the implications for policy of the perspective on rural change that has been presented are quite far-reaching. If the indicated trends are accepted the whole construction upon which the compensatory philosophy for rural areas has been built should be revised. The main strategies for rural areas should emphasize economies of diversification in relation to agricultural activities, local entrepreneurship, bottom-up approaches, small and medium enterprises, integration with the global economy, coherence with a segmented demand, maintaining flexibility, and competitiveness. This policy profile recalls closely the LEADER programme of the European Union. Perhaps it is this coincidence with current trends which lies at the basis of the success of that initiative.

Rural policy should not be agricultural policy in disguise. It could be asked why at the end of the 1980s the Common Agricultural Policy shifted towards a rural perspective. A plausible explanation is that *rural* policy has been invented in order to compensate farmers for the reduction in agricultural prices. This view would be supported by the constant confusion of agriculture and rurality that is found in many documents. But it is unduly restrictive: documents such as *The Future of Rural Society* (Commission, 1988) point towards a mixture of reasons, some more instrumental, others with a higher inspiration, but still in conflict.

Finally, it is ironic that those rural areas which benefited most from past agricultural policies, and became most productive and efficient, non-pluriactive and professional, now find themselves as in a more fragile position as reductions in price support and the current trends in favour of diversification give other areas, previously thought to be disadvantaged, a competitive edge.

# References

Bagnasco, A., 1977. *Tre Italie, La problematica territoriale dello sviluppo italiano*, Il Mulino, Bologna

Bagnasco, A., 1988, *La construzion e sociale del mercato*, Il Mulino, Bologna

Bagnasco, A., 1994, *Fatti sociali formati nello spazio*, Franco Angeli, Milano

Becattini, G. (ed.), 1987, *Mercato e froze locali. Il distretto industriale*, Il Mulino, Bologna

Becattini, G. and Bianchi, G., 1987, 'I distretti industriali nel dibattito sull' economia italiani' in Becattini, G. (ed.), *Mercato e forzelocali. Il distretto industriale*, Il Mulino, Bologna

Becattini, G. and Rullani, E., 1993, 'Sistema locale e mercato globale', *Economic e Politica Industriale*, 80: 25–48

Boyer, R., 1986, *La théorie de la régulation: une analyse critique*, Edition de la Découverte, Paris

Boyer, R., 1988, 'Alla ricerca di alternative al fordismo: gli anni Ottanta', *Stato e Mercato*, 24: 387–423

Brusco, S., 1989, *Piccole imprese e distretti industriali*, Rosenberg & Sellier, Torino

Chandler, A., 1994, *Dimensione e diversificazione. Le dinamiche del capitalismo industriale*, Il Mulino, Bologna

Commission of the European Communities, 1988, *The Future of Rural Society*, The Commission, Brussels

Dore, R., 1988, *Taking Japan Seriously: a Confucian perspective on leading economic issues*, Stanford University Press, Stanford

Organization for Economic Cooperation and Development, 1994, *Rural Indicators*, The Organization, Paris

Piore, M.J. and Sabel, C., 1987, *Le due vie dello sviluppo industriale. Produzione di massa e produzione flessibile*, Isedi, Milano

Porter, M.E., 1991, *Il vantaggio competitivo della nazioni*, A. Mondadori, Milan

Pyke, F., Becattini, G. and Sengenberger, W. (eds), 1990, *Industrial districts and inter-firm cooperation in Italy*, International Labour Office, Geneva

Salais, R. and Storper, M., 1993, *Le monde de production*, Edition de l'Ecole des Hautes Etudes en Sciences Sociales, Paris

Saraceno, E., 1994, 'Alternative readings of spatial differentiation: The rural versus the local economy approach in Italy', *European Review of Agricultural Economics*, **21** (3/4): 451–74

# Section IV: Canada

edited by
*Robert S. Dilley*

# Introduction
## Robert S. Dilley

Despite having a federal government and a prime minister continuing to enjoy very high public ratings a year after being elected, the long-term future of Canada has rarely looked less certain. The election of the separatist Parti Québecois (PQ) in that province, committed to pulling Québec out of Confederation, along with the presence of the separatist Bloc Québécois as official opposition in Ottawa, has left the whole country wondering what is going to happen next. The previous PQ administration (1976–85) of the late René Lévesque had tried, unsuccessfully, to get the people of the province to support *sovereignty-association*: a vague concept involving a large degree of autonomy while remaining within Canada. The current premier, Jacques Parizeau, is an unqualified separatist who wants Québec to be a fully independent country (while retaining many ties with Canada, including joint citizenship). However, many people voted for the PQ in search of reform rather than separation, and Parizeau will have to campaign hard to get a majority for leaving Canada.

If he and the PQ succeed (the referendum is supposedly to be held some time in 1995) there will be utter turmoil. There are no provisions in Confederation for secession. While there is a vocal minority in the rest of the country already crying 'good riddance', most Canadians are proud of their multicultural society and their image as people able to get along with anyone else. This self-image will suffer a catastrophic reverse if Québec should vote to leave. Further problems with possible secession concern the borders of an independent Québec. The PQ loudly trumpets the 'inviolability' of Québec within its present boundaries (what about the inviolability of Canada within *its* boundaries?) However, there is not even agreement about what those boundaries are. Québec claims most of Labrador as part of its territory (and publishes maps incorporating the claimed territory into the province) although it is in fact administered as part of the Province of Newfoundland, the boundary having been established by the Judicial Committee of the Privy Council in London in 1927. Much of northern Québec (the district of Ungava) was added to the province in 1912. Some constitutional experts argue that it should be returned if Québec secedes, while the almost entirely Indian and Inuit population of northern Québec has (to the fury of the PQ)

*Progress in Rural Policy and Planning*, Volume Five. Edited by Andrew W. Gilg
© 1995 Editors and contributors. Published 1995 by John Wiley & Sons Ltd.

expressed a strong desire to remain part of Canada should separation occur. A vote by the people of Québec to secede would merely be the *start* of national upheaval.

With attention firmly focused on national unity, it is easy to overlook other problems and developments. At the time of writing (December 1994) moves are underway to expand the North American Free Trade Area (Canada, Mexico and the USA) to the rest of the American continent, starting with Chile. The federal Liberal administration, when it was in opposition, had vehemently opposed the original free trade agreement with the USA and its expansion to Mexico, but now it sings the theme of continental free trade loud and clear. What effect this will have on rural Canada, notably the agricultural sector, remains to be seen.

One major issue that has occupied a good deal of the news *not* devoted to Québec has been the decline and possible death of the East Coast fishery. This volume contains a major chapter on this issue and its impact on the rural populations involved, especially those of Newfoundland, where fishing has been a major element of the island's culture since the first European settlement. So basic is this matter to the rural way of life that it seemed worth devoting the larger part of this section to it.

The annual review of trends and developments across Canada is a little shorter than last year, partly to accommodate Ommer's article and partly because I was away on leave in the UK most of the year and many regular correspondents took advantage of my absence to ignore my pleas for information (and the British press is apparently unaware that Canada even exists, except as a place the Queen is occasionally obliged to visit). Nonetheless, I found or was sent enough material to show that the environment continues to dominate much rural planning and policy making, and that there is a growing awareness of the role to be played in environmental planning by Canada's Aboriginal peoples.[1] Governments at different levels are paying more attention to Aboriginal needs and demands, and the Aboriginal populations themselves, especially the various First Nations, are becoming much more aware of their political power. This is a topic that might well prove worth closer investigation in a future issue.

---

[1] *Aboriginal* is the term increasingly preferred in Canada (replacing the more widespread but less accurate *native*) to describe people of Canadian Indian and Inuit ancestry. Early colonial administrations tended to negotiate solemn and binding treaties with the more organised Indian tribes in much the same way as the government in London negotiated with the nations of Europe. It is now usual to refer to those Indian tribes that signed official treaties with the federal government as *First Nations* and to accord them an increasingly high profile in the political process. Indians outside these Nations, and all Inuit, are *non-status* and wield much less political clout.

# 8 Canada: the rural scene
## Robert S. Dilley

## Farming

In May 1994 British Columbia introduced a number of amendments to its Agricultural Land Commission Act. The amendments were intended to strengthen the cooperation between the Provincial Agricultural Land Commission (PALC) and local governments by, among other things, making sure that local Community Plans are referred to the PALC for comment whenever they involve lands in the Agricultural Land Reserve (ALR). More power is to be given to local governments in certain areas. For instance, they will be able to refuse to authorise an application to proceed to PALC in any case where the land concerned is designated locally for farm use. The legislation also reconfirms all existing ALR boundaries, doing away with any uncertainty brought about by court challenges to the original designation process in the early 1970s. More information on this topic may be found in the chapter by Pierce and Séguin, in Volume Three, pp. 287–310.

## Rural planning

The federal Ministry of Small Communities and Rural Areas, created by the short-lived Campbell government (see Volume Four, pp. 243–4) has been retained in a modified form. Cross-Canada consultations, begun under the Conservatives to obtain input on the federal government's role in rural renewal, were continued under the Liberals. Assurances have been given that the sustainability of rural areas is a high priority of the new Minister of Agriculture. In February 1994 the establishment of a Rural Renewal Secretariat was announced, with offices (as before) in Winnipeg and Ottawa.

British Columbia's Commission on Resources and Environment (CORE, see Volume Three, p. 324) produced a report *Finding Common Ground: A Shared Vision for Land Use in British Columbia* which recommends that the provincial government adopt a new *Land Use Charter* and *Land Use Goals*. The *Charter* describes the fundamental principles of sustainable land use. It commits the government of British Columbia to protecting and restoring the

*Progress in Rural Policy and Planning*, Volume Five. Edited by Andrew W. Gilg
© 1995 Editors and contributors. Published 1995 by John Wiley & Sons Ltd.

environment and to securing a sound and prosperous economy for present and future generations. The *Charter* has four headings. Under *Environment* are included the need to conserve biological diversity; to anticipate and prevent adverse environmental impacts and to account for environmental and social costs in all decisions, on the 'polluter pays' principle. *Economy* encourages diversification; efficient use; making optimum use of the inherent capabilities of the land and ensuring the sustainable use of renewable resources. An undertaking in *Society* is that decisions will be made fairly and openly. This heading includes respect for the concerns of individuals and communities; a fair distribution of the costs and benefits of land-use decisions; and the promotion of opportunities to earn a living, obtain education and training and get social, cultural and recreational services. *Aboriginal* recognises Aboriginal title and the inherent rights of Aboriginal peoples to self-government.

The *Goals* set out the end results that British Columbians want from their use of land and water, under the headings: Resource Lands; Human Settlement; Transportation; Energy; Economic Development; Coastal and Marine; Cultural Heritage; Environment; Aboriginal Peoples; Recreation; and Protected Areas. Together, the *Charter* and the *Goals* will guide all provincial land use agencies and serve as advisory guidelines for local governments in making local land-use decisions.

Appendices to *Finding Common Ground* contain specific *Strategic Land Use Policies* detailing how government would achieve the desired ends defined by the *Goals*. These *Strategic Policies* provide general policy directions; a description of the factors to be taken into account by decision makers; direction regarding ways in which one set of issues should be integrated into consideration of other issues; instructions to take specific planning and management actions; and directions regarding the nature and content of guidelines, codes or standards that should be developed to achieve the *Goals*. Appendix I consists of a series of tables listing the *Goals* and the strategic policies that are designed to implement them. In some cases, specific *sustainability indicators* are included, so that quantifiable measures of success or failure can be used to assess how closely the *Goals* have been achieved. Columns are provided to show the figures for 1985, plus those expected in 1995, and those aimed for in 2000 and 2005 (for example, the figures given for the 'Percentage of solid waste diverted from landfills to recycling, reuse, or reduction' are: 35 in 1995; 50 in 2000; and 75 in 2005). Most columns are, however, blank and data are provided for only seven of over 120 indicators, although there must be readily available information for many others. Appendix II briefly describes current ministerial policies that already promote the same policies as the *Goals* and then identifies the 'Issues and Gaps' that need to be studied and discussed as part of a province-wide land-use strategy.

As noted in Volume Four on p. 252, British Columbia's CORE 'initiated planning negotiations in three regions: Vancouver Island; Cariboo-Chilcotin;

and Kootenay-Boundary'. The first of these has now been published: the *Vancouver Island Land Use Plan*. This is the first time that comprehensive land-use planning has been done at a regional scale in British Columbia. The Plan creates 23 new Protected Areas, connected by a network of Regionally Significant Lands that represent areas of major ecological, recreational and cultural significance. The remaining lands are divided into three broad categories: Multi-Resource Use Areas: Cultivation Use Areas and Settlement Land. The Plan also describes conditions that must be met to promote fair treatment of all people affected by its implementation. These conditions include an economic transition strategy to assist workers and resource-dependent communities, and consultations with First Nations to ensure protection of Aboriginal rights.

Also in British Columbia, the *Strategic Plan for the Fraser Basin Management Program 1993–98* is a product of the Fraser Basin Management Board; a body set up with representatives from the federal, provincial, local and First Nations governments, as well as representatives of environmental, industry and labour interests. The drainage basin of the Fraser River covers a quarter of the province, includes two-thirds of the population (expected to increase in total by 50 per cent in the next 25 years) and produces more than 80 per cent of the Gross Provincial Product. The *Plan* makes it clear that the Board is not duplicating the work done by CORE or the provincial Round Table on the Environment and the Economy (Volume Two, pp. 218–19) but rather is trying to coordinate and integrate the recommendations and activities of these agencies with all the other organisations, governmental and non-governmental, operating in the Basin. The *Plan* is replete with all the usual terminology and 'vision'; 'mission'; 'goals'; and 'sustainability' recur frequently. There are five identified goals: ecosystem conservation; resource management; community development; regional development; and institutional development. There seems little here that differs from CORE's document on *Finding Common Ground* other that its more local application. One innovative aspect of the publication is the use of sidebars to quote input from various public meetings. As one commentator in Abbotsford put it: 'Are we going to talk forever? We want action!'

In May 1994 the Ontario Minister of Municipal Affairs released a package of legislation (Bill 163), policy statements and administrative changes to reform the province's planning and development system. These changes are the end product of the more than three years' work carried out by the Commission on Planning and Development Reform in Ontario (Volume Two, p. 214; Volume Three, p. 324; and Volume Four, p. 246). The reforms are based on three principles:

(1) *Municipalities will be given greater control over the development process.* At present, planning authority normally rests with the province. Under the new system, municipal governments will make development decisions,

the province will set policy and the Ontario Municipal Board (OMB, the provincial planning appeals body) will adjudicate disputes.

(2) *The environment will be better protected by means of clear policy statements and legislative changes that integrate social, cultural, economic and environmental values.* Legislation will be changed to ensure that planning decisions are consistent with a new set of provincial policies covering natural heritage and ecosystems, community development and infrastructure, housing, agricultural land, energy and water conservation and mineral resources.

(3) *Red tape will be cut to make the planning process faster and more efficient, creating jobs in the construction industry and other sectors.* Legislative changes will set specific time frames for decision making by the province and by municipalities. Administrative changes (some already under way) will speed up decision making by municipalities, the OMB, and the Ministry of Municipal Affairs which will be the lead ministry for planning.

A twelve-member Advisory Task Force on Implementation has been set up, with four members from each of the Association of Municipalities of Ontario, the Ontario Environmental Network and the development industry. The mandate of the Task Force is to oversee all implementation guidelines being developed by the government on planning issues, and to advise on education and training requirements for the people who will be involved in the planning system.

## The environment

A good deal of progress has been made during 1993 in developing Canada's Ecological Science Centre (ESC) initiative. The ESC initiative arose from the 1990 Green Plan recommendation to establish 'a long-term state of the environment monitoring and assessment capability' (Volume Two, pp. 217–19). The aim of the ESC network is to provide a clearer understanding of ecosystems and how they work, as a base for economic and social decision making that is environmentally sensitive and sustainable. Ecological Science Centres are to comprise a number of 'anchor sites' in each of the country's fifteen major terrestrial and three marine ecozones, where long-term monitoring and research of ecosystems can be conducted. These sites may contain human-modified ecosystems such as farmland, commercial forests and urban areas as well as undisturbed natural areas. Each ESC is largely self-administering but forms part of a larger national network, with a national advisory committee to ensure common standards and facilitate cooperation and communication. So far, a wide array of partners — natural resources managers, ecological researchers and representatives from government agencies, non-governmental organisations, First Nations, universities and industry — have been working together in six ecozones, from Pacific Maritime to Boreal Shield. Most progress has been made in the Atlantic Maritime ecozone, where 'anchor sites' have been established in south-west Nova Scotia, centred on Kejimkujik National Park, and in southeast New Brunswick, centred on Fundy National Park. There is a newsletter, *ESC News*, which provides further information about the initiative: contact ESC, SOE Reporting, Fax (613) 941–9650.

A ten-minute video, *Earth's Harmony: An Argument for Changing the Way We Think*, uses a combination of graphics, animation and visual imagery to help describe ecosystems, their importance and how they function. The video contrasts two modes of thought: the reductionist and the holistic. The former, reducing a problem to its constituent parts and then dealing with each part individually, has long dominated our thinking and has taught us a great deal about the world around us. However, it is not a very good approach for understanding ecosystems as a whole. Here the holistic approach is more successful; looking at how all the parts fit into the greater whole; compelling us to reconstruct what we have taken apart and making us more aware of how integral a part we humans are in the environment. *Earth's Harmony* is aimed at a broad audience. Its short length allows it to serve as a catalyst for ecological discussions in schools, government, business and environmental interest groups. The video, along with another short presentation on Sustainable Development, is available for sale or rent through local offices of the National Film Board of Canada.

Statistics Canada's Environmental Information Service is extending its system of national accounts to include environmental assets and expenditures as well as a set of waste and pollution accounts. Work so far includes a set of oil and gas accounts for Alberta, an energy use account and a measure of greenhouse gas generation. Work in progress includes: extending the oil and gas accounts to the rest of Canada; accounts for other minerals; and a forestry account for Ontario showing timber stocks and changes by broad species, region, age and year. This information can be provided as maps or tables organised by drainage basin, ecozone or a variety of other geographical frameworks. The data can be aggregated according to user specification and combined with other data sets. Statistics Canada will be offering a CD-ROM containing data from their Environmental Information System as well as other related material.

Environment Canada's *State of the Environment (SOE) Bulletin* No. 94–4 is entitled *Climate Change* and, after a general introduction outlining the concept of global warming and its likely long-term effects, provides data in text and graphic form of Canada's production of the key environmental indicator, carbon dioxide ($CO_2$). Along with the use of fossil fuels (*SOE Bulletin* 94–3, summarised under 'Energy' below) emissions of $CO_2$ rose steadily until the mid-1970s, since when they have fluctuated but not risen significantly as total use of fossil fuels has levelled out and as more use has been made of natural gas in place of coal and oil. The principal producers of $CO_2$ are industry and power plants (49 per cent) and transport (29 per cent): especially gasoline- (petrol)-driven road vehicles.

The government of British Columbia tabled a revised Environmental Assessment Act in May 1994. The intent of the Act is to bring into a single process the assessment of major projects in energy, mining, industry, transport, aquaculture, food processing, water containment and diversion, waste disposal, and tourism. The proposed process guarantees significant public input, starting at an early stage. When a proposed project is on or near Aboriginal territory, First Nations peoples will be on the project committee. It is intended to coordinate the environmental assessment of a project with the application for relevant permits and licences.

## Forests

Canada has nearly 10 per cent of the world's forests and they are one of the country's major economic resources; accounting for over 700 000 jobs and more than $20 billion worth of exports. These data put into context the release of Forestry Canada's third State of the Environment report, *The State of Canada's Forests 1992*, in late 1993. The latest data show that harvesting increased by over 50 per cent from 1970 to 1990, though the economic recession caused something of a drop in 1991. In total, forests with commercial potential continue to grow at a faster rate than they are being harvested, though the rate of regeneration has slowed down somewhat and the area of forest depleted by natural causes (notably fire) continues to exceed the annual harvest. An encouraging note is sounded in the levels of reduction in pollutants — discharges of suspended solids, biochemical oxygen-demanding materials, dioxins and furans have all declined significantly in recent years — while the area of forest treated with insecticides and the use of herbicides have both declined. *The State of Canada's Forests 1992* is available, free of charge, from Forestry Canada, Policy and Economics Directorate, 351 St Joseph Blvd, Hull, Québec K1A 9Z9.

British Columbia's *Forest Renewal Plan* was introduced in 1994. An accompanying News Release states:

> For too many years, governments have taken the forests for granted. Too much has been cut and too little put back. We now face the prospect of a future with fewer trees and fewer jobs.

To counter this, the Forest Renewal Plan has as its goals to renew the land and keep the forests healthy, to ensure sustainable use and enjoyment of forests, to maintain the continued availability of forest jobs and to ensure the long-term stability of communities that rely on the forests. Over the next five years, an estimated $2 billion in new money is to be raised from stumpage rates (the money paid to the provincial government in proportion to the unit volume of wood harvested). This money, by law, is to be invested in the forests, the people who work in them and their communities. The investments will be managed by Forest Renewal BC; a partnership of government, the forest industry, workers, communities, First Nations and environmentalists. In the long run, the intention is to renew forests so that the next generation will have more trees to harvest on a sustainable basis. This will preserve more wildlife habitats and protect the environment. Money will be invested in forest worker training. The special interest of First Nations will be recognised by participation in joint ventures, forest worker training and participation in resource management programs. The impact of boom-and-bust commodity markets on forest communities will be reduced.

In April 1994 Ontario's Environmental Assessment (EA) Board released its long-awaited decision on the *Timber Management Class Environmental Assessment* submitted by the provincial Ministry of Natural Resources (MNR) (see the Chapter by MacCallum in Volume Two, pp. 204–6). This has been described (by counsel for the Ontario Forest Industries Association) as 'easily the longest, more extensive environmental assessment hearing in history'. It took almost six years to complete, heard from 552 witnesses in 411 days of hearings and generated over 70 000 pages of transcript. This was not a typical commission, designed to bury problems in words. Unlike most EA boards in Canada, decisions of the Ontario EA Board have legal effect unless overruled by the provincial cabinet within 28 days. From late May, therefore, the *Timber Management Class Environmental Assessment* became a legally binding code of forest practices for Ontario. Forest Practices Codes have been introduced in other provinces (e.g. British Columbia) but these have not undergone formal EA reviews. Ontario is therefore now at the forefront of forest management practices in Canada.

The conclusion of the Board was that Ontario's forests are, in general, being managed on a sustainable basis; despite evidence presented by conservation groups that commercial species are decreasing and that accelerated wood depletion will mean wood supply shortages, job losses and community instability. Essentially, MNR has been authorised to continue with its timber management undertaking, subject to over 100 conditions governing such activities as access, harvesting and renewal. One of the more controversial topics considered was the practice of clearcutting; removing every tree over a large area. The Board ruled that clearcutting *is* environmentally acceptable, and rejected requests to limit clearcutting to small areas only, on the grounds that this would make it impossible to return the boreal forest (which dominates Canada's forest landscape) to its natural pattern of large even-age stands. However, clearcutting is limited to 260 ha unless extra documentation supporting the need for a larger clearcut is provided.

The Board supported the continuing use of chemical herbicides and rejected calls to de-emphasise seeding and tree planting in favour of smaller cuts and natural regeneration. Overall, it has been healthy to have MNR's procedures subjected to public scrutiny, though there remains some concern whether enough has been done to ensure sustainability and biodiversity conservation.

Arising from this review, Ontario's Crown Land Sustainability Act (Bill 171) passed into law in December 1994. This Act places the responsibility for forest regeneration on the shoulders of logging and pulp companies. However, the passage of the Act has raised hackles among the province's Aboriginal population. The Grand Chief of the Nishnawbe-Aski Nation points out that in 1991 the premier of Ontario signed a statement of political relationship recognising the First Nations as distinct nations, to be consulted on a government-to-government basis. The Grand Chief argues, therefore, that industrial activities in forests traditionally used by Indians must first be approved by them. As well as their concerns about the forest environment, the First Nations want to be assured a share in any jobs or revenue arising from development of the forests.

## Energy

Environment Canada's *SOE Bulletin* No. 94–3 deals with *Energy Consumption* and looks at national consumption of energy and at the rate of fossil fuel consumption. Following a general introduction, there is a series of point-form comments highlighting key trends, and a number of graphs and pie charts. Total energy consumed tripled between 1958 and 1975, since when it has fluctuated but not risen significantly. A decline in energy consumption per dollar Gross Domestic Product suggests that the economy is becoming less energy-intensive. Renewable energy use is increasing slowly but steadily, and in 1992 made up 18 per cent of the total: two-thirds hydroelectric and one-third wood (mostly waste wood chips from pulp mills). Alternative energy sources, such as solar and wind power, make up less than one ten-thousandth of total consumption. Fossil fuels continue to provide most energy, though natural gas has recently replaced oil as the most important fossil fuel, with coal a distant third. The switch to natural gas benefits the environment, as it emits less carbon dioxide, sulphur dioxide and nitrogen oxides than oil or coal. Nuclear power now provides 11 per cent of the total. The bulletin ends with a look at 'Societal Response', outlining the federal government's new *Efficiency and Alternative Energy Program*, the *R-2000 Home Program*, teaching builders how to construct homes that use less energy and the Motor Vehicle Safety Act which have given the country some of the most stringent vehicle emission standards in the world.

Proponents of renewable energy will be pleased by developments in Alberta backed by the Alberta Energy Resources Conservation Board under the provisions of the provincial Small Power Research and Development Act of 1988. A major 'windfarm', with 25 eight-storey-high windmills, was completed in December 1993 on a high ridge near Pincher Creek in south-western Alberta. This was joined a few months later by a nearby 27-unit 'farm' jointly run by a Calgary developer and the local Peigan Nation. Individual private windmills have been commonplace over the Prairies for decades, but the only previous full-scale wind farm in Canada was built in 1987 at Cambridge Bay, Northwest Territories. So far, these developments have been met with enthusiasm from rural Albertans; and with none of the concern about landscape aesthetics and noise that have surrounded actual and proposed windfarm developments in Britain. Canada's huge open spaces and sparse population may well make it a much more suitable environment for such developments.

Alberta's Energy Resources Conservation Board (ERCB) is a quasi-judicial tribunal responsible, among other things, for reviewing proposed energy developments in the province. Environmental considerations have been part of its remit from the start, and these were reinforced and clarified in 1993 by the Energy Resources Conservation Act. In September 1994 ERCB caused a good deal of surprise by rejecting Amoco's application to drill an exploratory well in the Whaleback area of the province's Eastern Slopes. Amoco argued that the well was needed to evaluate the potentialities of its leases, and that an eventual discovery would produce significant economic benefits to the province. Opponents identified potential impacts on wildlife, on agricultural operation and on the rural lifestyle in general. Even safety for local residents was an issue, as there was a high probability that any hydrocarbons encountered would contain 'sour gas' ($H_2S$). The application was particularly controversial because the Whaleback, a scenic area and wildlife habitat, is said to be the largest remaining area of relatively undisturbed montane ecosystem in Alberta.

ERCB was concerned about deficiencies in Amoco's application and with its public consultation process. More significant in the long run was ERCB's interpretation of broader land-use issues. It noted that the application was inconsistent with the sub-regional Integrated Resources Plan (IRP), and that the Whaleback Ridge was a likely candidate for designation under *Special Places 2000*, the government of Alberta's proposed process for completing the protection of representative examples of Alberta's ecosystems by the year 2000. Environmentalists are celebrating a rare victory and the possible dawning of a new age in the regulation of the oil and gas industry in Alberta. Industry, however, considers the decision a 'betrayal' by an arm of government it had come to view as pro-business. There still remain issues to be clarified in ERCB's attitude towards IRPs and broader land-use issues. Until these are resolved there is going to be uncertainty in the oil and gas industry and also among conservationists and environmentalists.

## Fishing

Elsewhere in this section, Rosemary Ommer deals with the problems of Canada's East Coast fishery. Environment Canada's *SOE Bulletin* No. 94–5 considers *Sustaining Marine Resources: Pacific Herring Fish Stocks*. The bulletin points out that the West Coast herring are a resource that provides employment for thousands, a way of life for numerous coastal communities and contributes millions of dollars to the economy. Pacific herring are important as food (especially as traditional fare for Aboriginals), fishing bait and zoo food, and has a healthy export market (herring roe sells for $120–150 per kilogram in Japan). The bulletin describes and graphs the environmental indicator of spawning biomass for five Pacific coast locations from 1951 to 1993 and looks at the biology of the herring stocks and the threats to their habitat. It concludes that the most important factors in overall biomass are ocean temperature and the interrelated numbers of predators. The eight most abundant predatory fish harvested off the west coast of Vancouver Island are estimated to consume six times as much herring as the average annual fishery harvest.

## Parks and wildlife

The government of Nova Scotia is proposing to establish a comprehensive system of parks and protected areas in the province, based on a three-year inventory and evaluation of Crown Lands that remain intact as significant natural areas. The ideas are: to incorporate into the system representative examples of the province's typical natural landscapes and ecosystems; to include unique, rare or outstanding natural features; and to offer quality opportunities for wilderness recreation. Nova Scotia currently has 45 'Protected Areas' including four National Parks and five National Wildlife Management Areas totalling just over 160 000 ha. A further 31 areas, totalling some 287 000 ha, have been proposed as candidate protected areas. Together, existing and candidate areas would comprise about 8 per cent of the area of the province. A plan to set up this system is expected to be finalised and approved during 1995.

There is considerable debate over a 40 km stretch of the New Brunswick coast south and west of Fundy National Park. This is the longest stretch of mainland ocean front inaccessible by highway between Florida and the St Lawrence River. It is proposed to open the first 20 km of this coast to tourists by a two-lane parkway. The debate is not the usual one between developers and conservationists, but rather between proponents of different approaches to development and conservation. Much of the land is privately owned forest which contains mature stands of trees. With rising wood prices, many seem it as only a matter of time before the logging companies move in. A parkway, with the requirement of a substantial buffer to preserve views, would protect a large part of the area. The road itself would be designed to minimise its impact on the environment, following contours and avoiding sensitive areas. Trucks would be banned, and speeds kept low. Opponents argue that the estimated 1700 cars a day would destroy any vestiges of wilderness and drastically affect the habitat of animals like black bear and the eastern cougar.

Yukon's Department of Renewable Resources has found itself in trouble over its proposed wolf management plan in the Aishihik region, adjoining Kluane National Park near the Territory's southern border. The plan is part of an overall scheme to maintain a viable balance between populations of wolves, caribou and moose. Three Indian bands whose traditional hunting grounds include the Aishihik region complained in 1992 about diminishing moose and caribou numbers and increasing signs of wolf kills. After a reduction of sport hunting followed by a ban had done little to help, the territorial government culled the wolves to a level that studies suggested would maintain a balanced wildlife population. Now the Territory is under pressure from both sides, but has insisted it will not kill any more wolves unless numbers grow out of hand again.

## Rural resettlement

An encouraging story of rural settlement is told in the July/August 1994 edition of *Canadian Geographic*. After 60 years of being pushed off their land and out of their homes by more powerful interests, the Oujé-Bougoumou Cree have designed and built a permanent village. Originally they were a hunting tribe who spent summers together at Lake Chibougamou in north-central Québec. In 1927 a mining company cleared trees at a drilling site and destroyed dwellings. In 1936, Indian Affairs officials in Ottawa falsely declared the Chibougamou Crees to be part of the Mistassini Crees, 100 km to the north, and declared the whole Chibougamou area open to mineral exploration. Then followed decades of wandering, as site after site was rendered uninhabitable by mineral extraction, forest clearcutting and water pollution from mine effluent. Promises of band status were made but not kept, and a new village was bulldozed flat, with the authorisation of Indian Affairs, when a mining company needed sand. However, the group has survived both forced relocations and endless red tape from federal and Québec provincial officials. They now have their own land and a carefully designed village, some 40 km west of Lake Chibougamou. It incorporates modern technology, such as a central, sawdust-burning furnace to heat the whole village, and traditional designs, as in the semi-open meeting-house modelled after a traditional Cree dwelling. The story of the Oujé-Bougoumou provides an object lesson of a rural group holding on to its culture against all the odds.

## Rural studies

The University of Guelph has launched a new doctoral program in rural studies; the first of its kind in Canada. In developing the program, Guelph surveyed universities throughout North America to find graduate programs that might act as models. The survey found 76 programs in urban studies and just one specialising in rural studies. Guelph's initiative goes a little way to redressing the balance of academic interest between rural and urban communities. The Guelph program takes an interdisciplinary approach, looking at the social and economic dynamics of rural communities.

## Acknowledgements

I am grateful to the following individuals for providing information used in this report: Don Downe, Minister of Natural Resources, Nova Scotia; Curt Halen, Senior Planner, Ministry of Municipal Affairs, Ontario; Dennis Stephens, Rural Renewal Secretariat, Agriculture Canada and Brian Walisser, Director, Policy and Research, Ministry of Municipal Affairs, Recreation and Housing, British Columbia. Other items were taken from *State of the Environment Reporting*, the newsletter of Environment Canada; from *Canadian Geographic*, published by the Royal Canadian Geographical Society; from *Resources*, the newsletter of the Canadian Institute of Resources Law; and from the Thunder Bay *Chronicle-Journal* and *Times-News*. Interpretations and expressions of opinion are those of the Canadian editor.

# 9 Policy, fisheries management and development in rural Newfoundland

## Rosemary E. Ommer

## Introduction

Rural Newfoundland communities are currently in crisis as a result of the cod moratoria now in effect both off Newfoundland and in Atlantic Canadian waters as a whole. This is not just a regional, provincial or even national issue. It has significance for fishery-based communities the world over in so far as it is likely to be repeated again and again elsewhere unless (1) global management of fisheries becomes a reality rather than a rhetoric; (2) some way can be found to actually *manage* the world's fisheries as distinct from just talking about managing them or claiming to be already doing it; and (3) such management is designed to sustain communities of people, not just fish. This chapter seeks to discuss Newfoundland's fish industry in its broadest ramifications, and to identify key issues in fisheries management and coastal rural development. Newfoundland is used as an example and as a 'lens' through which the wider picture can be examined.

People have been fishing off the shores of this part of the world for nearly 400 years, but in permanent settlements on the island and in Labrador for only about half that time (Mannion, 1977). The first communities were established under the aegis of merchant fishing firms operating out of the United Kingdom, primarily the West Country, which spread around the coast over the course of the nineteenth century (Hiller and Ommer, 1990). Settlers operated a mixed economy in which the fishery was the centrepiece of a seasonal round involving hunting, trapping, woodswork, sealing and 'gardening' (preparation of vegetable gardens which provided the subsistence food requirements for settler families). Although the mix of productive activities varied from place to place according to the resources available, the logic of this way of life depended on a few basic premises which sustained existence here. Indeed, permanent settlement in Newfoundland became possible only when a way had been found to import the essentials of life

*Progress in Rural Policy and Planning*, Volume Five. Edited by Andrew W. Gilg.

which could not be produced locally at a reasonable cost. By the 1790s, the 'migratory' merchant fishery at Newfoundland (in which firms brought out indentured fishers and returned them to the United Kingdom after their period of service — usually two summers and one winter) had become uneconomic. What the merchant firms needed was a way to maintain their fishers *in situ* without having to pay them a year-round wage. The two interests were integrated when the merchant firms imported goods and exported fish for settlers, with the costs of the imported supplies being paid by the settlers in fish. For the rest of the year people hunted, trapped, cut wood, grew vegetables and fixed things. Some items were sold to the merchant, others were used purely as inputs to the household economy.

Over time there thus grew up a society of semi-independent households who fished under a barter arrangement for merchant firms in rural Newfoundland. 'Informally', they exploited the resources in the area which could not have sustained settlement on their own but did so in combination with the commercial merchant fishery (Ommer, 1990). This way of life resembled that of pre-industrial communities in many places around the North Atlantic rim and, indeed, in many non-industrial societies. In Newfoundland, it survived virtually undisturbed until the First World War (Ommer and Hong, 1990). Thereafter, a series of events weakened the old *modus vivendi* until, with Confederation with Canada in 1949, rural Newfoundland became a part of the industrialised First World (Ommer, 1994a).

Over the ensuing decades the fishery was modernised and compartmentalised until by 1970 it had three parts: an industrial offshore; a small-boat inshore; and a mid-shore which set uncomfortably between the other two. In 1977, the Canadian Government decreed a 200-nautical-mile exclusive fishing zone, and it also introduced policies favouring the technologically efficient offshore industry. The fishery entered a period of rapid expansion, with bigger fleets and an unprecedented expansion of fish-processing plants onshore, so that by 1990 the labour force of full- and part-time workers in the fishery amounted to nearly 44 000 people (Government of Canada, 1993: 157).

## The crisis

By 1990 the euphoria surrounding this expansion had collapsed along with the stocks of northern cod, to be replaced by panicked reappraisals. These took various forms whose expression was, in itself, instructive. In particular, scientists employed by the federal Department of Fisheries and Oceans (DFO) have been in an unenviable position over the last few years. Small and shrinking budgets ruled out adequate scrutiny of the waters off Eastern Canada and, as a result, their data suffered from large margins of error.

Naturally, the minister pressed them for information to help him set quotas which would serve the needs of all interests in the fishery. Among these was the urgent need of individual fishery workers to be employed for at least 10 weeks a year in order to qualify for unemployment insurance (UI), in a region where unemployment was severe (averaging 20 per cent in Newfoundland over these years). The provincial government, which licensed fish-processing plants, was likewise under pressure to keep as many of these functioning as possible. Profits and productivity could only fall. Investment in plants required a return, and that meant a lot of fish had to be caught, so the demands of industry kept quotas higher than was prudent. DFO calculated stocks and quotas using information from companies on their catches, rather than trusting entirely to their own data. That meant they confused 'effort' (what the fleet *could* catch) with the number of groundfish which could be caught without endangering the stocks (what the fleet *should* catch). They seriously overestimated biomass and set quotas well above advisable levels. Also, international fleets continued fishing beyond the 200-mile limit, the nations involved refusing to accept Canadian estimates of available stock, while diplomatic negotiations around reductions of their quotas moved so slowly as to be totally ineffective. Beyond that, fish may have been affected by environmental factors such as changes in water temperature and salinity, but these are not well understood. Almost no-one seems to have grasped the possibility that the great cod stocks of the North-west Atlantic could simply crash.

There were a few warning voices — in academia, in government advisory bodies, among marine scientists and in inshore communities — but no-one listened. The country was in temporary recession and some thought the region was permanently so. People were politically deaf to bad news about fish. Quotas were held low but level, stocks continued to decline, and plants continued to be licensed, as shown in Tables 9.1 and 9.2.

In 1992 the axe fell. Biological data had become unanswerable: the stocks were on the verge of collapse. They were clearly not commercially viable and perhaps not even biologically so. A moratorium on fishing in Canadian waters commenced with northern cod, and was then extended to the Gulf of St Lawrence stocks. Fishing communities all around Atlantic Canada were deprived of their major (sometimes their only) source of employment and industry, which was also the heart and soul of their 200-year-old culture and way of life. Industrially, the situation was critical; socially and economically it was disastrous; politically it was a nightmare.

The role of fishing in all of coastal Atlantic Canada is significant, but nowhere more so than in Newfoundland. In 1989, over 46 per cent of the people who derived some income from the fisheries of Atlantic Canada lived in Newfoundland and Labrador. These had the lowest gross fishing income and the highest amount of UI. The highest number of fish processors was also in that province. Moreover:

Table 9.1    *Canadian Atlantic groundfish allocations and catches for the years 1978–93*

| Year | All groundfish Allocation | Catch | All cod Allocation | Catch | Northern cod Allocation | Catch |
|---|---|---|---|---|---|---|
| 1978 | 472 | 535 | 204 | 271 | 100 | 102 |
| 1979 | 562 | 634 | 270 | 359 | 130 | 131 |
| 1980 | 705 | 615 | 353 | 400 | 155 | 147 |
| 1981 | 790 | 741 | 400 | 422 | 185 | 133 |
| 1982 | 924 | 775 | 490 | 508 | 215 | 211 |
| 1983 | 997 | 728 | 561 | 505 | 240 | 214 |
| 1984 | 1005 | 700 | 553 | 466 | 246 | 208 |
| 1985 | 1003 | 738 | 576 | 477 | 250 | 193 |
| 1986 | 973 | 748 | 530 | 475 | 250 | 207 |
| 1987 | 969 | 723 | 512 | 458 | 247 | 209 |
| 1988 | 985 | 688 | 523 | 461 | 266 | 245 |
| 1989 | 942 | 652 | 478 | 422 | 235 | 215 |
| 1990 | 812 | 604 | 408 | 384 | 197 | 188 |
| 1991 | 812 | 572 | 399 | 311 | 188 | 133 |
| 1992 | 808 | 418 | 333 | 182 | 120 | 21 |
| 1993 | 512 | na | 121 | na | Moratorium | |

Entries are thousands of tonnes

*Source*: Government of Canada (1993, Appendix C, DFO Resource Allocation, p. 124).

Table 9.2    *The number of fish-processing plants registered with DFO for Canada's Atlantic provinces in the years 1981–91*

| Year | Total | Nfld | PEI | NS | NB | Québec |
|---|---|---|---|---|---|---|
| 1981 | 700 | 225 | 42 | 213 | 101 | 119 |
| 1982 | 685 | 215 | 42 | 225 | 114 | 89 |
| 1983 | 670 | 205 | 43 | 237 | 127 | 58 |
| 1984 | 724 | 212 | 48 | 256 | 140 | 68 |
| 1985 | 788 | 213 | 55 | 278 | 163 | 79 |
| 1986 | 840 | 228 | 60 | 327 | 135 | 90 |
| 1987 | 890 | 244 | 56 | 307 | 177 | 106 |
| 1988 | 991 | 252 | 65 | 343 | 207 | 124 |
| 1989 | 975 | 256 | 76 | 345 | 187 | 111 |
| 1990 | 1018 | 268 | 75 | 347 | 190 | 138 |
| 1991 | 1063 | 281 | 75 | 348 | 194 | 165 |

Nfld = Newfoundland, PEI = Prince Edward Island, NS = Nova Scotia, NB = New Brunswick

*Source*: Government of Canada (1993, Appendix C, DFO Inspection Data, p. 146).

For many . . . income . . . earned from employment outside the fishery, often by additional workers in a fishing family, is combined with income from other sources. The evidence suggests that all sources of income are important for most of those who fish for a living. It is also clear that access to UI, one of the major sources of non-fishing income, has created strong linkages between the industry and the income security system (Hollett and May, 1993, p. 185).

When one the adds the fact that 76 per cent of the fisheries workforce had less than high-school education in 1986, anyone can see how hard it would be for these people to find something else to do, given their poor education (Carter, 1993).

So far, the discussion has been limited to a purely economic mindset. If we consider the social dimension, we see that a whole culture is in danger, a way of life, a tradition, a sense of belonging that is rooted in the fishery, particularly that of the inshore. Inshore fishers in Newfoundland remain to some degree a non-industrial people, rooted in community and place, and in a way of life that has little to do with the urban, industrial wage worker's nine-to-five routine (Ommer, 1994a). Furthermore, this is not usually sufficiently appreciated (if it is recognised at all) by policy-making bodies, with the result that government and the rural Newfoundlanders they serve are often at cross-purposes and bewildered by, on the one hand, the apparent intransigence of the populace to make what are seen as necessary adjustments to their lives and, on the other, the inadequacy of those in power to help. The end result, as Schrank *et al.* (1992) summarises it, is that:

Traditional industries often coexist uneasily with modern economies . . . . [T]hey have often become uneconomic and subsidized, employing more people than they can support at the prevailing standard of living. Nevertheless, because of the traditions engendered in the community, the cultural ambience of the people, and the apparent lack of economically viable alternative employment, the uneconomic traditional industry tends to remain in place in its traditional form for far longer than is warranted on purely economic grounds (pp. 335–6).

They add that '[T]he problem is a common one both in less developed countries and in many areas of the more industrially developed world'.

In Newfoundland, with the moratoria, all this has come to a head. The stakeholders involved are now facing a crisis. A full year into the moratoria, the offshore industry has begun to make adjustments. The big companies have sold many of their vessels to Third World nations and they have begun to buy fish from other suppliers (like Russians, who fish the Barents Sea). The multinational large companies are still solvent and may just pull out of these waters entirely. Politically and economically, the picture remains very confused, and the reason for that seems to be that there is no agreed-upon assessment of what has happened and therefore no unified vision of what needs to be done.

## Explanations and solutions

The various fragmented views of the situation can be encapsulated as follows. From the perspective of federal marine scientists (as presented, at a recent international conference on the global management of fisheries resources, by a representative of DFO with 30 years of service behind him) the problem is a complex one of attempting to untangle the strands of environmental and resource variability under circumstances in which the department has found itself more often concerned with the management of people rather than fish. In this view, all sectors of the industry hold some measure of responsibility for the collapse of the stocks since small (between 45 and 65 feet) vessels as well as larger boats can do a great deal of damage. Industry will, therefore, have to become an active and willing partner in conservation, fishers will have to be included on research vessels, and a major restructuring of the industry will have to occur (Beckett, 1994).

From the perspective of the offshore sector, the fishery could be made viable by concentrating the industry in a few large ports, with a few high-technology vessels, a few coordinated and high-technology processing plants, and a skilled, numerically small but highly paid workforce. This view was expressed in an oral presentation at the *Defining the Reality* conference held at St John's, Newfoundland, in March 1993. The perception is shared in its essential components by the federal government Task Force on the Atlantic Fisheries, chaired by Richard Cashin, one-time President of the Newfoundland Fishermen, Food and Allied Workers Union (Government of Canada, 1993). The Task Force investigated at some length the complex issues of location of the future industry and the need for education and training of fishers so that they might become recognised as a 'professional' workforce. The union appears to concur with this analysis, although with an understandable degree of discomfort.

The provincial government thinks about it from the angle of employment, and invokes the need to diversify the province's rural economy. There is also a related issue of fisheries jurisdiction involved here, centred on the long-standing debate in Newfoundland about federal/provincial management of the fishery. The federal government in Canada has jurisdiction over the fishery to 200 miles offshore, while the province has control of fish-plant licences and freshwater fisheries. There are some who feel that truly joint management of the marine fisheries would have prevented the present crisis and will be essential to rational restructuring when/if the stocks recover. Tightly connected to these matters is the issue of UI (a federal concern) and welfare (a provincial matter) — a constant subtext to debates on the fishery ever since Newfoundland's Confederation with Canada in 1949, and now underlined by the recent national and provincial recessions and ongoing budgetary deficit issues as well as the parlous state of the Newfoundland economy which has rarely looked secure for most of the twentieth century (Ommer, 1994a).

It is clear, as evidenced at the 1993 *Defining the Reality* conference, that rural people know that they are facing irrevocable change. They know that not all communities are going to survive, and they realise that the reckoning will come somewhere. They are braced, in a way never seen before, for radical change. Their most common response to the current situation is a plea to be informed, and quickly, about which fish plants are going to close for good, which communities are marked for extinction, and what their inhabitants are going to be able to do instead. There is a great deal of fear and anger in fishing communities. A radio show on the topic (*Fishery Forum V* on CBC, 17 July 1994) contained interviews with fishers from the south coast of the province who spoke passionately of their dilemma. 'I am scared to get up too early in the morning,' said one. 'I might crack if I have to do nothing for a long day.' Financial concerns are pressing, too. As one man said: how is a family of husband, wife and five children to pay loans on the boat and truck and deal with other expenses on $380 per week? The comment that 'I wish they'd tell us who'll be in and who won't so we could plan something' has been made again and again. There is anger as well. Referring to urban Canadian perceptions and prejudices about fishers' life-styles and their supposed refusal to move elsewhere to seek employment, one man said 'People who say we're lazy should try getting up at five in the morning in January and going out to fish'. He added, 'Tell the politicians they're not moving *me* off this Rock. *They* can go, if they want to, but I'm not.' (Though people do actually move. There has been an absolute and net loss of population to the province in the last decade, much of it coming from the better-educated labour force cohorts.)

Among all the uncertainty, one thing is very clear. The province is actually at a crucial — perhaps the most crucial — period in its history right now. The initial shock is over. The reality has been accepted: in the stark words of one fisher, 'Let's face it — we've caught them all'. That is no exaggeration. The most recent statistics on cod stocks in Newfoundland waters show the lowest biomass estimates on record and the report states flatly that 'recovery cannot be anticipated before at least ten years'. The same is true for most other groundfish stocks in the region (Department, 1994). And so people are now prepared, as never before, for change . . . and that is both tragic and a window of opportunity for the future. The question is what the nature of that change needs to be, and it involves coming to grips with the landward as well as the marine problems of Newfoundland development.

## Rural diversification

In recent years (since 1986), the need to diversify the rural economy of the province in ways which are compatible with the scale, skills and lifestyle of rural Newfoundland communities has been gaining some measure of acceptance in political and industrial sectors, as a result of the work done by a provincial 'Economic Recovery Commission' which was put in place to explore the economic options for Newfoundland and Labrador in the 1990s (Government of Newfoundland, 1986). The Commission recognises the need for fishery restructuring and also for state financial support for individuals during that restructuring, given that any process involving long-term economic redirection will not happen overnight and will involve substantial outlay of funds in income support, retraining, development funds and the like. It also clearly recognises the difficulties of creating alternative landward development for communities which will not be able to be involved in fishing in the future.

The Commission is, however, handicapped (as is the province and, indeed, the nation) by the fact that, given current economic conditions, the political instinct (and, to some degree, the fiscal necessity) is to provide quick-fix solutions which distract from, and may even deter, the efforts to put more lasting, but slower, major changes into effect. In a 1994 public lecture, the Chair of the Commission cited pessimism in government officials in the face of the collapse of the 'old [resource-based and government-expansionist] economy' as clouding the prospects of a better, more diversified future (House, 1994). He also spoke of the evils of Newfoundland's long-term dependency on federal government transfers with the concomitant vulnerability to changes that can take place outside of the province's control, particularly in the face of governmental fiscal stringencies and the continuing escalation of public sector debt. The good news, he claimed, lay in the existing basis for the development of a 'new economy' which could be created around a whole range of small industries operating at the provincial scale, many of which would not be fishery-related. The groundwork for this was already laid: the 'knowledge intensity' of provincial firms, especially outside the fishery, approached the national average as did the education level of the workforce, again outside the fishery; post-secondary institutions were beginning to adjust to the requirements of the new economy where most new job creation was also developing; there was a shift in this direction in both new and old industry associations and community-based economic agencies; and labour was also beginning to look more seriously at the long term. What is needed now, he insisted, is careful management of the transition and rapid expedition of the whole process, and a focused vision, incorporating landward and seaward initiatives for future growth, instead of uncoordinated, piecemeal and reactive attempts to merely survive yet another crisis.

Unfortunately, that is easier to say than to do, particularly for those parts of the province where the feasibility of the 'new economy' is far from obvious in those small isolated settlements: which have depended almost exclusively on the fishery; which have poor transportation linkages to even regional centres; and where alternative resources, either natural or of a service nature, would appear to be in very short supply. Nonetheless, there are few, if any, other options open to a region which has been ill-served by 'top-down' mega-project salvations in the past. Local people know that. Increasingly they are frustrated by government short-term thinking and want to participate in the planning process. For example, one Labradorian commented bitterly on a radio program (*Fishery Forum VI*, CBC, 24 July 1994) that at present the government is taking the impetus for shaping the future away from places and putting the emphasis instead on just 'surviving for today'. There is a growing international recognition of the value of the strengths and ideas of local people and the small-scale regional diversification strategies that they usually suggest and support. As a recent article expressed it:

> Once it might have been called self help, now it's more sophisticated and, because of its increasing success in the developing world, it's more thought-through . . . it's time-consuming, exploratory, mediatory rather than confrontational and has no guarantee of results (Vidal, 1994, p. 25).

This is precisely the kind of 'bottom-up' development that local people in Newfoundland are espousing at present, for both the land and the sea (always assuming that some marine resources may have survived the depredations of the last few years). The work of Daniel Pauly on tropical artisanal fisheries speaks to this issue very clearly. He (and many others like him) argues that such fisheries have, in the past, been sustainable, being shaped by the needs of communities rather than the requirements of sophisticated 'efficient' technology — though one may question calling a technology which does serious damage to the marine environment 'efficient' — and with limits on entry organised around the clan or group, rather like the Newfoundland traditional fisheries which were organised by community (see also Dahl, 1988; Jentoft, 1989; Sinclair, 1990). In recent years, however, tropical artisanal fishing effort has increased dangerously and sustainability has become problematic. These fisheries now catch an amount equal to half of that of the developed world, making them very important fisheries globally. This escalation of effort is partly the result of disturbance in the agricultural sectors of such nations' economies which, as they have modernised, have released excess labour into the fisheries. Pauly sees the necessity, in the Third World, of creating coastal management structures in which fishing communities will have a voice, learning to work *with* fisheries scientists and governmental agencies as they jointly search for a way to restructure their landward and seaward economies. He notes, in passing, that this pattern is usually Third, not First, World, but makes an exception for Atlantic Canadian fishers (Pauly, 1994). In Newfoundland, encouragingly, part of what he advocates is starting to happen. Joint work by fishers and marine biologists is already under way as academic social scientists, DFO biologists, academic ocean scientists and fishers explore ways to integrate fishers' knowledge of local ecology with that of natural scientists in order to achieve a more coordinated approach to marine resource management (Neis *et al.*, 1994). But many problems remain and they are, in both Pauly's case and that of Newfoundland, very similar to those described in the *Guardian Weekly*, the principal one being that small-scale, bottom-up development thinking

> can . . . send authorities into a spin because the ordinary rules may not apply and it implies a switch of power towards people. Moreover, it seldom fits into budgetary plans . . . .

However, the article continues:

> But around the world this 'bottom up' work is producing astonishing results and . . . it may be heralding one of the more significant shifts in development thinking in 30 years. Its strength is that it applies equally to First and Third Worlds, binds together environment and development, and draws ethical and social issues into the development process — something treated by governments and agencies as exercise in economics alone (Vidal, 1994, p. 25).

The appearance of articles like this in a major First World newspaper is important, if only because it signals the beginning of the kind of media awareness that can eventually promote change in governmental thinking. Certainly, to many working in academia and in other areas where rural development issues are well known, the kinds of results that are reviewed in that newspaper article are already very familiar. Sadly, those involved in the daily grind of governing usually regard bottom-up small-scale development strategies as: esoteric or romantic; unpractical; not cognisant of the practical difficulties; theoretical day-dreaming and the like, notwithstanding, current success stories. Traditional economists too, who are so influential in governmental thinking on these issues, are likewise inclined to dismiss efforts to adopt 'small is beautiful' strategies as inefficient at best, impossible at worst. Academics are often responsible to some degree for such labels. Some of the literature on bottom-up development is heavily layered with jargon and is offputting for the more traditionally minded by virtue of this, and by the use of writing styles which smack of proselytising and ideological righteousness. This is sad, because there is much of value in some of the work currently being undertaken, especially by those attempting to produce a new and credible theoretical economics of sustainability (for example, Jannson *et al.*, 1994).

In North America at least, there is a growing openness among those concerned with fisheries management to the idea that social scientists are needed by natural scientists if there is to be an effective response to the potential global depletion of fish stocks. Even better, there are signs that many issues which have hitherto been seen as separate are beginning to be discussed as part of a 'package' that must be coherently grasped and dealt with. These are: managing fish (the biological problem), managing the industry (the quota/ licensing and the technology/investment return problem), developing adequate international regulations (the straddling stock problem), managing the business (the adequate returns and shifting markets problem), and helping the un[der]employed fisher (the socio-economic problem). While these issues are common to fisheries management the world over, it is at the regional scale that their lineaments are most easily grasped and the linkages between component parts most readily identified. So also are the stumbling blocks that stand in the way of solutions. In a place like Newfoundland, these are the realities of local politics, local fears, local defensiveness and local shortages of funds.

## Global significance

There are a few central points that come clearly out of the Newfoundland situation, which can usefully inform the global picture. Perhaps foremost is the insistence by some key players that one should not use fishery policy to achieve social objectives. This is perhaps true, but what is unfortunate is that too often a false corollary is drawn, and that fishery policy must therefore ignore the socio-economic circumstances and consequences of stock management. This is nonsense and, what is more, dishonest, in that any discussion of quotas and licences is actually a discussion about wealth distribution which is, by definition, political. In New Zealand fishing quotas have been privatised so that they can be traded, and banks there regularly accept the possession of an ITQ (individual transferable quota) as a legal asset. New Zealand champions ITQs in global fisheries management conferences and debates. It is, moreover, a false dichotomy to set the welfare of fishing communities against that of fish stocks. Such thinking subscribes to an either/ or way of reasoning which is counterproductive, especially in such a complicated area as fisheries management.

We would do better to face the thorny issues head-on and acknowledge that fisheries in Newfoundland and around the world are changing radically; that a resource sector which until recently operated for the most part at the 'hunter–gatherer' level of development (although a technologically sophisticated version of that) is now experiencing the growing pains of incipient industrialisation, with all that entails in terms of management, technology, property rights and wealth distribution.

If we consider the fisheries in this way, a number of issues are immediately clarified. The path to industrialisation has involved an increasing ability to expand the human capacity to exploit resources. Put in current fisheries jargon, it 'expands effort' through the application of more and more effective technology to the 'gathering' (harvesting) process. Eventually a point is reached which is akin to the Malthusian 'precipice', where the carrying capacity of the natural resource to sustain its predators (for that is what we, as consumers of fish, are) is reached, and the mortality of the resource rises sharply. That is the point we have now reached in Atlantic Canada. Malthus wrote in these stark terms about human populations at the beginning of the nineteenth century, when increased population in Britain seemed to be outstripping the capacity of the land to sustain it. What would follow unless something was done, he said, were the brutal controls of war, vice and famine on population expansion (Pauly, 1994). But Malthus was wrong, at least for that time, because human ingenuity came into play, and the agricultural and industrial revolutions expanded the production of food and manufactures. They did so partly through the creation of new technologies, partly through a concomitant expansion of trade and colonization, and partly through the opening up of the vast prairies of the United States, Canada, Australia and Argentina as transportation and then refrigeration technologies made that possible. The end point of that whole process is the global economy in which we live today.

The inevitable extension of this process has to be that we now face a world in which further expansion through redistribution of what are now beginning to be seen as finite natural resources (trade) is close to its limits. The work on 'ecological footprints' now being done by Rees and Wackernagel recasts the well-known concept of expanding urban resource hinterlands, and their extension through trade, in terms of the unrecognised ecological impact such forces have had on the carrying capacity of global natural resources. This serves to remind us that, while we can refine and develop our trade patterns, there are no new lands to discover, no brave new worlds on this planet to provide us with a way to avoid the consequences of stretching the appropriation of our 'natural capital requirements' beyond their carrying capacity, and thus escaping once more the dire predictions of Malthus (Rees and Wackernagel, 1994). There is no place left to hide, in effect. The old antagonism between ecology and economy needs to be squarely faced and our problems of resource sustainability acknowledged and solved. In the fisheries, this means coming to grips with the biological implications of the technologies we now use, and seeking to develop beyond the hunter–gatherer stage into an agricultural stage which is, of course, what aquaculture is.

Involved in all this will inevitably be a debate around property rights. ITQs and other distributional tools in fact find their historical parallels in the enclosure which went hand in glove with the development of agriculture in the past. The enclosures, of course, enclosed land which had previously been held in common, and they involved impoverishment for rural people as a consequence of the expropriation of land previously held by communities under 'rights of ancient possession'. The same process now appears to be under way with sea-based resources. Indeed, in Newfoundland now there is some speculation that the courts will shortly have to deal with challenges to the ban on the 'recreational' (i.e. daily food) fishery on the basis of historic rights (*The Express*, 20 July 1994). The 'tragedy of the commons' — a phrase which has become a buzzword in current debates on entry into the fishery — extends to the destruction of communities and livelihoods as well as resources, it would appear, although Garrett Hardin, who coined the expression, was thinking only in biological terms (Hardin, 1968). There is evidence that once again it is rural people and small communities who are suffering in the expropriation of marine resource rights as the industrialisation of the fishery proceeds apace. That is not to say that management is not needed, nor even that some of the marine 'rights of ancient possession' will not need to be altered. It is to say that those involved in the management planning process need to be aware that fishing communities perceived such 'rights' as existing and that planners will now need to act with caution, fairness and sensitivity as they set about restructuring.

Unfortunately, the assumption behind the way in which Hardin's work is currently being used is that Western society always operates in an individualistic manner, and that it is the pursuit of individual profit that drives entry into the fishery and ensures that effort will continue until it becomes unprofitable, or beyond, if subsidies create the conditions which so permit. This is the context in which the Scott Gordon rationale for fisheries management is applied, using total allowable catches, quotas and the like, with government acting as the 'unified directing power' which Scott Gordon saw as essential to the wise prosecution of marine resources (Gordon, 1954). Governments, therefore, should manage entry into the industry in such a way as to halt the individualistic drive to catch every last profitable fish regardless of resource mortality. The mortality of 'fishing communities' is, however, nowhere to be found in his analysis.

There are two other elements of the Scott Gordon analysis (beyond the omission just noted) which need more careful consideration. The first — an erroneous one — is that, unlike what Hardin theorised, not all Western societies do (or did) operate on a capitalist, individualist basis. Many small fishing communities, those of Newfoundland among them, managed their fishing resources at the community level and, moreover, have a long history of being ecologically aware (Thornton, 1980). The second — which has potentially positive consequences — is that, while Gordon and Hardin are correct that commons become tragedies when the cooperative property element fails, or when individualist wealth accumulation enters and disrupts the communal system (as has happened in Newfoundland and in the artisanal tropical fisheries of which Pauly writes), they both assume that capitalism is always individualistic at the level of the business person or firm. But it appears that that may not always be the case. Indeed, there is increasing evidence that, under 'high capitalism', industrial firms behave as *groups*, as communities of special interest. This is the principle of the cartel. But an interesting new version of that is coming into being in the fisheries as, for example, with the technologically sophisticated fishing firms of New Zealand, who are negotiating with their government for the allocation of ITQs to them as a group with shared interests. They are behaving, in a sense, tribally. But they do not have any history of being ecologically aware or sensitive — quite the reverse — and that does constitute a threat to conservation. But it is also possible to see it as evolution towards better management of the fisheries of the future, if it can be believed that the lessons of overcapitalised fleets have been learned.

In short, whether science and industry like it or not, any conservation approaches in the future will depend on politics, on political will and on prevailing ideologies at the level of the nation-state and the international community (if there is one). That is where the essential bottleneck to future sound fisheries management really lies. The problem is not ultimately that of 'open access' any more, although that has been a comforting way of blaming a given structure rather interest groups. Even in the so-called 'open' deep-sea, there is access control because there are considerable entry costs. The real issue is the management of wealth distribution, and that is always political, whether expressed as an employment or an investment or even a development issue.

The logic goes something like this. To maximise jobs (which, through wages, mean wealth) or profits, or both, there has been a race for fish. This has meant overcapitalisation of large fleets and of some small ones too, and overextension of fishing rights (however achieved). That has led to biological overshoot through waste (large by-catches) and overfishing, which has been facilitated at least in part by uncertain scientific biomass information (including possible contributory climatic factors). This in turn meant that the biological warnings did not come either soon enough or loud enough or convincingly enough to prevent it. That has led to political, economic and social stress which has put pressure on politicians and industry, and led to an increase in risk-prone decision making with respect to 'effort'. The end result has been the situation in which we find ourselves. Were we, however, to start the logic sequence in another place – say, with conservation — we would immediately have to move to conservative stock estimates and the fact is that we do not really know enough to be able to make those securely. We would therefore have to move to seriously reduced quotas of one form or another, and that would immediately lead to the politics of wealth distribution, which in turn requires us to consider jobs and economic development, which does not translate merely into quotas for firms. It also translates into the survival of communities. And it will, of course, mean looking at landward as well as marine opportunities — a restructuring process that will return to outport Newfoundland (and elsewhere) the kind of integration of land and sea activities that it used to have and now needs to find again in twenty-first-century terms and with twenty-first-century appropriate technologies.

## Conclusion

The management of fish is ultimately about the society in which the fishery is prosecuted. It is self-deception to think we can talk about the sustainability of fish communities if we do not, at one and the same time, also talk about the sustainability of fish*ing* communities. In the industrial parts of the First World that may not be immediately apparent, for urban centres are somewhat remote from the ecosystems which sustain them (a point made very nicely by the concept of the 'ecological footprint') but it is nonetheless true. Indeed, in effecting the shift of the principal factor of production from land to capital, the Industrial Revolution signalled a loss of contact with, sensitivity to, and awareness of the physical environment (Ommer, 1994b). First World civilization thought it had conquered nature in the process of 'proving' Malthus wrong. But it now appears that ecology may be catching up with its old antagonist, industrial economy. We may have lost not only the Dodo and the Great Auk but also the ordinary North Atlantic codfish — the raw material for the fish and chips or fish fingers that urban dwellers are used to getting in the greasy spoon café down the road. At the end of the day it may not only be the rural cultures of places like Newfoundland but the lifestyles of urban dwellers, too, which will be affected by the damage we are inflicting on our marine ecosystems.

## Acknowledgements

The work for this chapter has been funded by the Tricouncil Initiative of the Social Sciences and Humanities Research Council of Canada, the Natural Sciences and Engineering Council of Canada and the Medical Research Council of Canada, and by the Institute for Social and Economic Research, Memorial University of Newfoundland. I am grateful to Roxanne Millan, Peter Sinclair and Stuart Pierson for their comments on earlier drafts of this paper.

## References

Beckett, J., 1994, 'Trends in marine fisheries management in Canada', Paper presented at the *Fisheries Management; Global Trends Conference*, University of Washington, Seattle
Carter, B.A., 1993, 'Employment in the Newfoundland and Labrador fishery', in Storey, K. (ed.), *Defining the Reality*, ISER Books, St John's, pp. 132–75
Dahl, C., 1988, 'Traditional marine tenure: a basis for artisanal fisheries management', *Marine Policy*, 14: 40–48
Department of Fisheries and Oceans, 1994, *Stocks Status Report, 94/4*, Department of Fisheries and Oceans, Atlantic Fisheries Division, Atlantic Stock Assessment Secretariat, Dartmouth, Nova Scotia

*The Express* (St John's), 1994, 'First fisherman nabbed breaking cod-jigging ban', Lead article, 20 July, 1

Gordon, H.S., 1954, 'The economic theory of a common-property resource: the fishery', *Journal of Political Economy*, 62: 124–42

Government of Canada, 1993, *Charting a New Course: Towards the Fishery of the Future: Report of the Task Force on Incomes and Adjustment in the Atlantic Fishery*, Government Canada, Ottawa

Government of Newfoundland, 1986, *Building on Our Strengths: Final Report of the Royal Commission on Employment and Unemployment*, Government Newfoundland, St John's

Hardin, G., 1968, 'The tragedy of the commons', *Science*, 162: 1243–8

Hiller, J.K. and Ommer, R.E., 1990, *Historical Background Report on the Canada France Maritime Boundary Arbitration*, Government of Canada, Departments of Justice and External Affairs, Ottawa

Hollett, A. and May, D., 1993, 'An overview of incomes of workers in the Newfoundland and Labrador fishing industry', in Storey, K. (ed.), *Defining the Reality*, ISER Books, St John's pp. 176–95

House, J.D., 1994, 'The ugly, the bad and the good: the Newfoundland economy and its prospects', seminar at the Institute of Social and Economic Research, St Johns, 2 February

Jannson, A.M., Hammer, M., Folke, C. and Constanza, R. (eds), 1994, *Investing in Natural Capital: the Ecological Economics Approach to Sustainability*, Island Press, Washington, DC

Mannion, J.J., 1977, *The Peopling of Newfoundland: Essays in Historical Geography*, ISER Books, Memorial University of Newfoundland

Neis, B., Felt, R., Haedrich, R. and Schneider, D., 1994, 'An interdisciplinary methodology for collecting and integrating fishers' ecological knowledge into resource management', paper presented at *Fifth International Symposium on Society and Resource Management*, Fort Collins, Colorado

Ommer, R.E., 1990, 'Introduction', in Ommer, R.E. (ed.), *Merchant Credit and Labour Strategies in Historical Perspective*, Acadiensis Press, Fredericton, New Brunswick, pp. 14–15

Ommer, R.E., 1994a, 'One hundred years of fishery crises in Newfoundland', *Acadiensis*, 23: 3–20

Ommer, R.E., 1994b, 'Environmental history', paper presented at the *Social Sciences and the Environment Conference*, Ottawa

Ommer, R.E. and Hong, R., 1990, 'The Newfoundland fisheries: the crisis years, 1914–1937', Paper presented at the *Atlantic Canada Studies Conference*, Orono, Maine

Pauly, D., 1994, 'Trends in artisanal fisheries: science, management and human implications, fisheries management', paper presented at *Global Trends Conference*, University of Washington, Seattle

Rees, W.A. and Wackernagel, M., 1994, 'Ecological footprints and appropriated carrying capacity: measuring the natural capital requirements of the human economy', in Jannson, A.M. *et al.* (eds), *Investing in Natural Capital: the Ecological Economics Approach to Sustainability*, Island Press, Washington, DC

Schrank, W.E., Roy, N., Ommer, R. and Skoda, B., 1992, 'An inshore fishery: a commercially viable industry or an employer of last resort', *Ocean Development and International Law*, 23: 335–6

Sinclair, P.R., 1990, 'Fisheries management and problems of social justice', *MAST*, 3: 30–47

Thornton, P.A., 1980, *Dynamic Equilibrium: Population, Ecology and Economy in the Strait of Belle Isle*, PhD dissertation, University of Aberdeen

Vidal, J., 1994, 'Local know-how works wonders', *Guardian Weekly: Environment section*, 17 July: 25

# Section V:
# Australasia

edited by
*Geoff McDonald*

# Introduction
## Geoff McDonald

The Australasian Section covers that part of the world from Papua New Guinea through to Tonga including Australia, New Zealand and the South West Pacific. Unfortunately, the editors have been unable to locate a correspondent to review developments in the Melanesian and Polynesian Islands and would be delighted to hear from interested contributors. The Section in this volume is limited to Australia and New Zealand.

At the national level in Australia, the farming community and rural policy makers are struggling to cope with yet another drought, seemingly the worst this century, and to maintain some semblance of order in rural adjustment policy in the face of natural disaster. For the first time for over a century Australia may be forced to import grain. After five consecutive years of negative farm income, the national government has revised the capital asset test applying to social security to allow farmers to receive unemployment payments and other benefits without having to sell their farms.

Michael Taylor investigates the financial performance of broadacre farming at the regional level in Australia by assembling a unique integrated data set of economic and environmental variables and searching for relationships using a multiple regression model. He is able to document the widespread and general decline in profitability in Australian agriculture, due mostly to low commodity prices on world markets. He discovers interesting and worrying evidence to show that national performance in broad acre cropping has been maintained by expanding production into more marginal regions and environments. If he were able to assess the extent to which yields of Australia's grain were dependent on fertiliser applications to mask decline in natural soil fertility, the conclusions would be even more telling in terms of sustainable agriculture, or rather the lack of it.

Roche, Le Heron and Pomeroy examine the realigning of food power in New Zealand via changes to the electoral system to a German-style mixed member proportional system of elections. New Zealand's major food processor has recognised organically grown fruit and vegetable as niche exports and is expanding output considerably. New Zealand's central government is changing its approach to rural development focused on community involvement and self-help to attract job opportunities into rural regions.

*Progress in Rural Policy and Planning*, Volume Five. Edited by Andrew W. Gilg
© 1995 Editors and contributors. Published 1995 by John Wiley & Sons Ltd.

In order to accommodate these two chapters, the Annual Review of rural planning has been held over until Volume Six. It is hoped to include an extended two-year review in this volume.

## Note

Please note that in Volume Four of PIRPAP the following changes should be made.

i)  Page 301, last line – should read 'compared with about 6.3 per cent *or* 16 million hectares held in the conservation estate.'
ii) Page 306, line 11 – should read – 'Carbon *did have* a proper contract, which had been renewed . . .'

# 10 Realigning food power in New Zealand

## Michael Roche, Richard Le Heron and Anne Pomeroy

Last year's report covered producer board marketing, afforestation and the Resource Management Act 1991. A year on, these three areas remain significant but other new issues have emerged including: the realignment of rural power via changes to the electoral system; the expansion and regulation of organic fruit and vegetable exports; and central government's approach to rural development.

## Repainting the political landscape

The November 1993 election saw the National Party retain office, with a single-seat majority. Of greater long-term significance for the political landscape was the referendum held in conjunction with the election in which 54 per cent voted to replace First Past the Post (FPP) with a German-style Mixed Member Proportional (MMP) system of government. An earlier referendum in September 1992 had gained sufficient support for the government to offer the FPP–MMP choice in 1993.

The weaknesses of a FPP 'Westminster' system has been explored in the New Zealand context by Johnston (1976, 1979). More recently, the report of the Royal Commission on Electoral Reform Towards a Better Democracy recommended an increase in the number of MPs, changes to electoral administration, and the introduction of an MMP voting system. Policies of public and private sector restructuring and deregulation initiated by successive Labour (1984–90) and National (1990–) governments contributed to a deeply felt public cynicism about politicians' motives and integrity which by 1993 crystallised into referendum support for MMP. The referendum results deserve careful analysis to discern different patterns of support for MMP and FPP across north-south and rural–urban electoral cleavages, but space precludes such discussion here. However, some other implications of

*Progress in Rural Policy and Planning*, Volume Five. Edited by Andrew W. Gilg
© 1995 Editors and contributors. Published 1995 by John Wiley & Sons Ltd.

MMP are briefly highlighted. The first of these relates to the new electoral boundaries under MMP and the second to the realignments in rural politics brought about by the change.

Under the MMP system the 95 general and four Maori electorates (constituencies/seats) will be reduced to about 60 general and five Maori electorates.[1] The population of general electorates will increase from about 35 000 to 50 000. An additional 55 seats will be allocated on a proportional basis. These MPs will neither be directly voted for nor have a specific electoral constituency. Parties can enter Parliament by winning 5 per cent of the vote which will entitle them to five seats. This part of the MMP system has attracted the most attention to date, especially the means by which parties will fill their non-electorate list seats (e.g. Hawke, 1993).

The Representative Commission charged with drawing up the new electoral boundaries completed draft maps by 15 August 1994. They will initially hear submissions from political parties and later from the public with the boundaries being finalised by April 1995. In redrawing the boundaries a number of National Party seats, including Far North, Wallace, Wairarapa, Pahiatua and perhaps Albany, are expected to disappear (Benseman, 1994). This poses some difficulties for the National party especially as three Cabinet Ministers (including the Deputy PM) could be affected. Historically the National Party has embraced farmers' interests (Gustafson, 1986), but the change to MMP means more than the disappearance of some rural and mixed electorates to the National Party's disadvantage.

A detailed commentary must, however, await the publication of the draft boundaries. McRobie (1993), however, has produced a set of maps anticipating the Representation Commission's work. He reduces the South Island to 15 general electorates with 44 in the North Island. It is possible to compare these with the old FPP electorates by using James and McRobie's (1990) analysis of electorates which identified, *inter alia*, rural and mixed electorates. An approximation of these categories to the possible MMP electorates as shown in Table 10.1 reveals little overall change in the number of mixed electorates, i.e. those with rural and small town populations and economies, but a halving of the number of rural electorates in both islands.

If McRobie's boundaries are accurate, how will rural voters fare under MMP? There is agreement that MMP will herald an era of coalition government. This could have a number of possibilities for rural New Zealand. The politics of cooperation may favour the largely urban population which, in turn, will doubtless produce tensions, since the export basis of the national economy is likely to remain oriented towards agriculture. On the other hand, a new 'Country Party' may emerge to represent the political and economic aspirations of some in rural New Zealand. The ideological basis of parties may be most clearly displayed by 'list' MPs. In all events, both

---

[1]  Calculated on the basis of number of voters registered on the Maori rolls.

Table 10.1 *Rural and mixed electorates FPP and MMP*

| | Rural electorates | | Mixed Electorates | |
|---|---|---|---|---|
| | North Island | South Island | North Island | South Island |
| FPP boundaries | 14 | 7 | 9 | 4 |
| MMP boundaries | 7 | 3 | 8 | 4 |

Compiled from McRobie (1993) and James and McRobie (1990)

National and Labour have been implicated in the restructuring and deregulation of the 1980s to the extent that they no longer represent some of their traditional core values.

## G(r)o(w)ing organic

New geographies of consumption and production are at work in global Agri-Commodity Production Chains (ACPC). Baker and Crosbie (1994) identify an emerging market segment unsatisfied with food safety guarantees and especially concerned about pesticides. Organic farming typically, apart from mainstream production, has been in a small but significant way drawn into the emerging global food system. The international trade in organic food is variously estimated at from $75 million to $300 million, New Zealand's contribution being $3 million (Minister of Conservation, 1994). Tradenz estimates, however, that New Zealand producers might be able to contribute anything from $8 million to $125 million (Minister of Conservation, 1994). The driving force in New Zealand has been a commercial vegetable processing company – Watties Frozen Foods (WFF) (owned by J.W. Heinz[2] since 1992) — moving in response to customer demand (Ladds, 1992).

WFF, which is New Zealand's largest frozen food processor, announced during the year that it was now seriously committed to exploring processed organic vegetables (after the first trials in 1991). Currently WFF is one of only three export organic processors worldwide and these crops constitute less than 1 per cent of the international frozen food output (*Grocers Review*, 1993; *Export News*, 1993). The potential export market is large and estimated at $NZ 5 billion for both Europe and the USA by 1995 (Chalmers, 1993). WFF production is, however, only at present several hundred tonnes of peas, beans, carrots and sweet corn destined for Japanese and US markets. The Japanese market has been opened by the recent reduction of trade barriers for agricultural products and processed foods. Attention has been

---

[2] Heinz President, Tony O'Reilly, identified New Zealand as an 'Environmental Oasis' in 1993 consistent with the move into Asian markets and the interest in organics (Head, 1993).

focused on Japanese beef imports but vegetables and seaweeds constituted 12.25 per cent of food items consumed by Japanese households in 1990, second only to fish and seafood (Riethmuller, 1994). New Zealand organic growers have realised prices 30 per cent higher than 'orthodox' methods, even though average yields have been lower than for conventional systems.[3]

WFF has advertised for additional contract growers in Canterbury, Horowhenua, Manawatu, Hawkes Bay and Gisborne. Harvesting and processing requirements will lead to distinctive geographies of organic production. A minimum area of 4 ha is required for cost-effective harvesting. WFF has also purchased frame and tine weeders for hiring to growers. Processing must occur within 4–8 hours' travel time (approximately 45 km under current harvesting/transport regimes) from the Hawkes Bay canneries (Edmond, 1993). The combination of organic growing and more conventional processing for export is perhaps a contradictory relationship — it does, however, reveal the flexibility export food processors can display in creating new market niches for their products.

Joint research with a Crown Research Institute is also under way to establish biological control programmes (for example, against tomato fruit worm) and to produce mildew-resistant pea cultivars (*The Press*, 1994a). Government scientists at Hort Research also produced 200 cartons of Royal Gala apples for export via the Apple and Pear Marketing Board (APMB). The consignment sold in the UK was grown on organic lines to Biogro standards. Other varieties — Harvard Red, Oregon Red and Braeburn — are also being grown to organic standards. The APMB hopes to be able to capitalise on an important niche market where they have not previously been able to supply (*The Press*, 1994b). Certification from the New Zealand Biological Producers and Consumers Council (Biogro Standards) is required for would-be producers (*Export News*, 1993). However, additional government certification is proposed with the Ministry of Agriculture and Fisheries (MAF) likely to be required to promote this system in order to meet EC standards which will incorporate a quality management audit (*New Zealand Commercial Grower*, 1994).

Organic vegetable production is illustrative not only of new niche markets for processors but also of the influence of consumer preferences in Europe, the USA and Japan on complete agricultural production, harvesting and transport systems in New Zealand. New regulatory codes are aimed at guaranteeing food standards and quality (Wills and Harris, 1994).

*Michael Roche and Richard Le Heron*

---

[3] Organic farming advocates claim that when all costs and benefits are included their system is more economically viable than conventional farming systems.

## Reconstructing rural development

With some notable exceptions (Pomeroy, 1994), the history of rural development in New Zealand has been firmly focused on agriculture. Official policy in the post-war era until the mid-1980s, promoted agricultural production with particular emphasis on technical issues. Planning policy complemented agricultural science's concern to preserve land resources for agriculture by attempting to bar residential development on the most fertile soils. During the 1960s the government sponsored a series of conferences aimed at industry development, starting with the Agriculture Development Conference (ADC) in 1963 and culminating with the National Development Council in 1969. The outcome of these conferences was the introduction of macro- and micro-economic thinking into agricultural policy development (Nightingale, 1992), but the emphasis remained with agriculture.

Within agriculture the Manpower Working Party of the ADC had highlighted the loss of workers from the industry as long ago as 1964. It was recognised that there was little information on the social factors affecting farm 'manpower' supply and rural population dynamics. This concern resulted in an official review of rural social conditions — with particular reference to farm 'manpower' (Lloyd, 1974), and the broadening of economic analysis to include social issues. Social research effort was stepped up, particularly in the universities, while the 1977 Town and Country Planning Act provided an opportunity for more flexible planning and gave recognition to 'social' issues as a purpose for planning. However, there was little official recognition of the need for social policy analysis (Pomeroy, 1994) or of the need to broaden the development focus from agriculture to a broad multi-industry approach.

Instead, by the end of the 1970s the thrust of the government's rural policy had moved into major intervention to support deteriorating economic conditions, rising unemployment and regional disparities. Primary industry was seen as the only area of potential for economic growth and a system of subsidies was put in place to ensure returns above the guaranteed minimum price while overseas prices were low (Nightingale, 1992). Falling agricultural commodity prices in the late 1970s and early 1980s meant that this policy created a considerable financial debt on the state.

With the change in government in 1984, the policy of rural adjustment through price support, concessionary interest rates and fertiliser subsidies was abruptly turned around as the rural and national economies were opened to market forces. Regional development programmes (under the Department of Trade and Industry, now Commerce) which had supported the establishment of manufacturing industry in regional and rural service centres also came to an end in the mid-1980s, and resources were moved to support a new business development programme.

During the 1970s and 1980s there was considerable lobbying by rural organisations to persuade the bureaucracy and politicians to recognise that many of their decisions were having an adverse effect on rural livelihoods and living conditions. This was assisted by a tacit acknowledgement that agricultural policy should recognise the dependence of the agricultural industries on a servicing infrastructure which in turn depended on a mix of primary, secondary and tertiary industries for economic survival.

The government's response in 1991 was to expand MAF's work on the technical and economic performance of the farm and agri-business sector to include an official social policy perspective. This includes providing advice to the government on: the access of rural people to basic services; changes in the rural and farm socio-economic and demographic structure; and the social factors contributing to agricultural viability and sustainability. While there is no official rural development policy, the government has targeted resources towards programmes which directly or indirectly assist local economic development. These programmes are provided by several government departments including: the Community Employment Group; Commerce's Business Development programme; Internal Affairs' Link Centres (one-stop-information-shops); Maori Development; and MAF's facilitation work. The last is an explicit policy shift which recognises the need to consider the farm sector as part of the whole rural economy, and particularly the integration of the farm household with a wide range of industries (such as tourism), and acceptance of the range of resources within rural areas which can be utilised sustainably for economic gain.

A major new feature of 1990s rural policy is the inclusion of stakeholders in the policy development process. To be successful, development programmes require effective community ownership of ideas and the activities which flow from them. By taking a partnership rather than controlling approach, government is recognising that the future of farming, rural service centres, and communities is largely in the hands of the individual farmers, business proprietors, employees and people who live and work there.

While it is true that the job markets of most rural areas are restricted, there are considerable opportunities for niche market developments and there are many resources in rural areas which have not yet been tapped. By working cooperatively, flexibly and creatively to plan and develop strategies for using local resources, local people are able to introduce new business and job opportunities into their area. Many are doing this. The government's role is to help foster more local initiative, and to sponsor training in areas such as business and financial planning (for those with basic, but not advanced skills) in a non-patronising way.

The application of economic principles may leave equity issues inadequately addressed. As the government gets better at providing communities with opportunities to express what they want for themselves, there may be increasing need for resource allocation to take account of the communities' perceptions of what they prefer. Governments at all levels can play a role in this by assisting key people and groups to find solutions in order to make things happen. Central government, in partnership with local governments and the rural sector, can facilitate change and business development through providing an effective institutional framework, leadership and policy direction, for example in relation to infrastructure, incentives/penalties, seed funding, information, training, and improved service delivery; and through removing impediments to growth. Government can assist in establishing the policy context and economic climate that will encourage private sector and community initiative, change attitudes and motivate people.

*Ann Pomeroy*

*Authors' note*

Any views or opinions expressed here do not necessarily represent the official view of MAF.

# References

Baker, G.A. and Crosbie, P.J., 1994, 'Consumer preference for food supply attributes: a market segment approach', *Agribusiness*, 10(4): 319–24
Benseman, P., 1994, 'Natural boundaries: key to security', *Evening Standard*, 4 May
Chalmers, H., 1993, 'Organic-vegetable growers counted as exports increase', *The Press*, 12 November
Edmond, K., 1993, 'Watties still try on organic crops', *Horticultural News*, December
*Export News*, 1993, 'Watties backs organic vegetable growing', 13 December: 13
*Grocers Review*, 1993, 'Committed to major programmes in 1994', 72(12): 56
Gustafson, B., 1986, *The First 50 years: a History of the New Zealand National Party*, Reed Methuen, Auckland
Hawke, G.R. (ed.), 1993, *Changing Politics? The Electoral Referendum 1993*, Institute of Policy Studies, Victoria University of Wellington, Wellington
Head, W., 1993, 'Heinz head sees New Zealand as "environmental oasis"', *Export News*, 1 November: 1
James, C. and McRobie, A., 1990, *Changes? The 1990 Election*, Allen and Unwin, Wellington
Johnston, R.J., 1976, 'Spatial structure, pluriactivity systems and electoral bias', *Canadian Geographer*, 20: 310–28
Johnston, R.J., 1979, *Politics, Electoral and Spatial Systems: an Essay in Political Geography*, Clarendon, Oxford
Kennedy, E., 1994, 'Bid demand for organic vegetables', *The Dominion*, 10 June
Ladds, J., 1992, 'Wattie takes organic crop', *The Press*, 25 March
Lloyd, D.W., 1974, *A Preliminary Review of Rural Social Conditions with Particular Reference to the Manpower Position of Farms*, Agriculture Production Council, Wellington
McRobie, A., 1993, 'Electoral districts under MMP: one possible electoral map', in

Hawke, G.R. (ed.), *Changing Politics? The Electoral Referendum 1993*, Institute of Policy Studies, Victoria University of Wellington, Wellington: 9–42

Minister of Conservation, 1994, *Press statement: Organic farming systems face an image problem*, 13 July

*New Zealand Commercial Grower*, 1994, 'Organic standards', February/March: 17

Nightingale, T., 1992, *White Collars and Gumboots: A History of the Ministry of Agriculture and Fisheries 1892–1992*, Dunmore Press, Palmerston North

Pomeroy, A., 1994, 'Review of Rural Resources Unit Programme: integrated rural development', MAF Public Information Paper No. 8, Wellington

Riethmuller, P., 1994, 'Food processors, retailers and restaurants: their place in the Japanese food sector', Australia–Japan Research Centre, Pacific Economic paper, No. 230

Royal Commission on Electoral Reform, 1986, *Appendices to the Journals of the House of Representatives* (AJHR) H3

*The Press*, 1994a, 'Path to organic growing easier', 21 January

*The Press*, 1994b, 'Canty organic apples exported to UK', 22 April

Wills, I. and Harris, J., 1994, 'Government versus private quality guarantees for Australian food exports', *Australian Journal of Agricultural Economics*, 38(1): 77–92

# 11 International protectionism and regional dimensions of changing farm financial performance in Australian broadacre cropping

*Michael Taylor*

## Introduction

Australian broadacre agriculture is in crisis. During the 1980s broadacre farms producing cereals, sheep and beef began to experience severe financial difficulties that have persisted and intensified into the 1990s. These difficulties have arisen from a combination of factors associated with drought, rising costs and falling prices. Drought, especially localised drought, is common in Australia (Jeans, 1986; Williams, 1967), and Australian farmers' costs have risen persistently in the post-war years, while the prices they receive for their production have fallen in response to high levels of world output and protectionism (Miller, 1987; Knopke and Harris, 1991).

Protectionism abroad, which undermines traditional agricultural markets, is currently Australia's greatest rural trading problem. Protectionist measures have been imposed in Europe and parts of Asia principally in the name of food security, while protectionist measures have been introduced in the USA purportedly to combat the protectionism of others. The villain of the piece has been cast as the European Union's Common Agricultural Policy (CAP) that, in less than ten years, has transformed Europe from a net importer to a net exporter of temperate-zone agricultural products. The USA has cried foul, but has used its own Export Enhancement Program (EEP) as much to ameliorate its domestic farming problems, brought on by an overvalued dollar and massive farm debt, as to play the principled white knight fighting trading wrongs. At the same time, only minor inroads have been made into the highly protected markets for agricultural products in Asia, especially in South Korea and Japan.

Increasingly, Australia is being left on the outer in world markets, as the global economy appears inexorably to split into three trading blocs. In

*Progress in Rural Policy and Planning*, Volume Five. Edited by Andrew W. Gilg

consequence, Australian agriculture is now undergoing a process of massive structural adjustment as a low-cost (though some would say simply 'cheap' (Webber, 1990)) agricultural system battles for continued economic relevance in a rapidly changing world order.

At the regional scale within Australia, the impact of protectionism and recession on farm financial performance has always been far from uniform. Farms in some regions appear, on average, to perform better in comparative financial terms than those in other regions, even within the same sector of broadacre production (Australian, 1991; Department of Local, 1987). This variability reflects, in part, the patterning of physical and environmental conditions in Australia (Parvey, 1988; Williams, 1990; Jeans, 1986; Barr and Cary, 1992). However, management too is by no means a uniform input between regions. The family farm that dominates broadacre production has goals, decision-making strategies and capital structures significantly shaped by family cycle conditions, inheritance and succession issues, non-commercial attitudes to risk, and other personal, social and cultural pressures. The variable combination of these physical and environmental conditions, family business and management conditions, and rapidly changing trading circumstances associated with a continuing cost/price squeeze, divide Australia into sets of relatively better-performing and worse-performing regions of broadacre agricultural production.

In this chapter the magnitude of these regional variations in Australian broadacre farm financial performance is examined against the background of worsening cost and market conditions and intensifying recession during the 1980s and early 1990s. The analysis focuses on the broadacre livestock and cropping sectors in the period 1994–5 to 1990–1 using data for 33 regions from the Australian Bureau of Agriculture and Resource Economics (ABARE) from which three indices of farm performance have been developed: average farm cash operating surplus; rate of return on labour; and the ratio of debt to equity.

## Agriculture subsidy and market distortion

Australian broadacre agriculture is relatively unprotected and trade exposed. At least two thirds of rural production by value is exported, while, at only 9 per cent, the effective rate of protection of Australian agriculture is low by international standards (Curran et al., 1990).

Protectionism and subsidy plague world agricultural production and trade, with the European Union(EU) and its CAP leading the way. Through the CAP, 60 per cent of the value added in EU agriculture comes from subsidies, with the result that world prices for temperate-zone agricultural commodities have been lowered by as much as 16 per cent (Australian, 1988; Bureau, 1985). From Australia's perspective the impact on cereals and beef prices has been dramatic. Subsidies have converted the EU from a net importer to a net exporter of wheat and coarse grains, with the import of stock feed into Europe further swelling the pool of grain to be exported (Roberts *et al.*, 1992). At the same time, the CAP has pushed beef prices in member countries well above world price levels (Tsolakis and Sheales, 1990), and stocks have been sold on to world markets at highly subsidised prices. However, the impact on Australia has been ameliorated somewhat by issues of foot and mouth disease-free status and the Andriessen Assurance on EU exports to the Asia Pacific region (Harris and Dickson, 1990). Overall, the impact on Australian broadcare farming has been massive. The value of Australian cereals production has been depressed by as much as 30 per cent, and it has been estimated that the cost to Australia has been in excess of A$ 1 billion a year (Australian, 1988; Bureau, 1985).

The impact of the CAP has been compounded by US action in the form of the Export Enhancement Program (EEP) which was introduced in the mid-1980s in retaliation for unfair competition in export markets, especially from the European Union (Love and Clark, 1990; Roberts and Whish-Wilson, 1993; Roberts and Love, 1989; Roberts *et al.*, 1991). The situation was, however, somewhat more complicated and was equally attributable to conditions in the USA in the first half of the 1980s and the need to ease US farmers through a period of crisis. For Australia, the result has been subsidised US grain sales into traditional markets in Asia and the Middle East and a reorientation of Australian domestic production to other previously less profitable areas of broadcare production. Although Australian losses from the EEP have varied from year to year, the National Farmers Federation has estimated that all US farm subsidies cut Australian farm incomes by over A$700 million, between 1985 and 1991 according to an article in the *Financial Review* newspaper on 2 January 1992.

Protectionism elsewhere in the world, notably in East and South-east Asia, and especially in Japan and South Korea, has heightened the problem for Australian agricultural exports by denying them access to one of the few alternative markets. The Japanese situation is particularly difficult deriving again from a national policy of food security. In the late 1980s it was estimated that Japanese agricultural protection, alone, had reduced the world meat trade by up to 27 per cent and had reduced world meat prices by as much as 7 per cent (Reithmuller *et al.*, 1988).

## Costs, debt and restructuring

The cumulative impact of increased overseas protection on Australian broadcare agriculture has been to intensify an existing cost price squeeze. The reactions of farmers to these falling prices, radically changed market conditions and rising costs have included:

Table 11.1   *The changing structure of broadacre farm input costs*\*

| | Annual average 1977–8 to 1979–80 | | Annual average 1987–8 to 1989–90 | | 1990–91 estimated | |
|---|---|---|---|---|---|---|
| | $m | % | $m | % | $m | % |
| Fuel | 907 | 5.0 | 1 070 | 5.1 | 1 110 | 5.7 |
| Fertiliser | 951 | 5.2 | 1 112 | 5.3 | 890 | 4.6 |
| Chemicals | 420 | 2.3 | 884 | 4.2 | 820 | 4.2 |
| Seeds & fodders | 2 064 | 11.4 | 2 115 | 10.2 | 1 950 | 9.7 |
| Marketing | 2 605 | 14.4 | 2 196 | 10.6 | 2 180 | 11.2 |
| Repairs & maintenance | 1 425 | 7.9 | 1 753 | 8.4 | 1 545 | 7.2 |
| Other costs\*\* | 2 405 | 13.3 | 2 879 | 13.8 | 3 585 | 18.4 |
| Wages | 1 968 | 10.9 | 2 585 | 12.4 | 2 660 | 13.7 |
| Interest paid | 1 050 | 5.8 | 2 839 | 13.6 | 2 450 | 12.6 |
| Capital expenditure | 4 325 | 23.9 | 3 363 | 16.2 | 2 250 | 11.6 |
| Total | 18 118 | 100.0 | 20 796 | 100.0 | 19 440 | 100.0 |

\* In 1990–91 dollars
\*\* Sharp rise in the wool tax in 1990–91

*Source*: Knopke and Harris (1991, p. 231)

- Farm amalgamation and farm enlargement
- Increased farm production (by some 15 per cent in the 1980s)
- The introduction of new technology (with environmental spin-offs in the form of clearing remnant vegetation and exacerbating salinity problems)
- Product diversification
- Adoption of new land management practices
- The restructuring of costs.

Throughout the post-war years, the prices received by Australian farmers have increased less rapidly than the prices they have had to pay. This cost price squeeze intensified in the 1980s and early 1990s. While farm receipts have declined in these years in the face of increasing protection overseas, costs of all inputs have also continued to increase (Knopke, 1990). Consequently, pressure on farm receipts has translated rapidly and directly into attempts by farmers to reduce or at least contain their costs.

Two factors have complicated recent attempts to contain farm costs. First, as shown in Table 11.1, relative prices of inputs have changed and, second, changing production methods have significantly affected patterns of input use. Expenditure on new plant and equipment halved in the 1980s, while expenditure on repairs increased. At the same time, expenditure on chemicals and fertilisers grew rapidly with the adoption of minimum tillage land management practices; a development that also helped to contain fuel costs (Barr and Cary, 1992; Knopke, 1990; Knopke and Harris, 1991). However, a particularly important change was the major increase in interest payments associated with rapidly rising levels of farm debt (Peterson *et al.*, 1991; Knopke and Harris, 1991).

Indeed, four sets of pressures have combined to drive some sections of Australian broadacre farming deeply into debt, and in some cases towards bankruptcy (Backhouse *et al.*, 1989; Peterson *et al.*, 1991). Real interest rates rose from 0.6 per cent in 1979–80 to 13.5 per cent in 1989–90; although they fell in both 1992 and 1993. Bank deregulation in 1983 made borrowing by farmers much easier, but also reduced their concessional sources of borrowing. The synchronised downturn in rural commodity prices left farmers with no room to alter production strategies, and lifecycle pressures forced some farmers to borrow heavily to enlarge their holdings to provide for their children and the next generation of farming families. As a result, real average interest payments rose during the 1980s from approximately A$6700 to approximately A$14 000: from 5.6 per cent of farm costs to 13.6 per cent, but falling back to 12.6 per cent of farm costs in the early 1990s (Tucker *et al.*, 1990). Nevertheless, at the beginning of the 1990s, farm debt in Australia was still 16 per cent higher than it had been a decade earlier, although it had peaked in 1986–7.

The nature of the cost price squeeze that has affected Australian farmers in the past 15 years has also affected the ability of farmers to substitute one form of broadacre production for another. From the early 1980s to the present, prices for Australian wheat and beef have fallen almost continuously. For wool, however, the story has been very different. While wool prices remained reasonably stable during the first half of the 1980s, they rose dramatically in the later years of the decade only to slump disastrously between 1989 and 1991. It was quite logical therefore that, especially in Australia's wheat-belt regions, wool production should have been substituted for wheat production in the late 1980s. However, between 1989 and 1991 the international market for wool turned sour, and falls in the wool price began to parallel falls in the prices of both beef and wheat (Connolly *et al.*, 1990;Roper *et al.*, 1990). Now, for the first time, price falls in all areas of broadacre farming became synchronised, leaving Australia's broadacre farmers with no scope to substitute one form of production for another in order to maintain their incomes.

However, to suggest that the income and debt situation on all broadacre farms is uniformly bleak is quite wrong. Restructuring for some farms has been extremely successful and this shows up in rapidly widening income differentials on broadacre farms (Australian, 1991, Lawrence, 1987; Taylor, 1993). Using ABARE data on farm incomes, overall there has been an increase of 79 per cent in the six years 1984–5 to 1989–90 in the income gap between Australia's best-performing and worst-performing broadacre farms. Moreover, even debt is not a major problem on a significant number of broadacre farms, and ABARE has suggested that while 31 per cent of farms are completely without debt, a further 10 per cent hold assets in excess of their debts (Peterson *et al.*, 1991).

## The differential regional impact of rural recession: the example of broadacre cropping

It would be wrong to infer that all regions in Australia have been affected to the same extent by the impact of the current, largely protection-induced recession in Australian agriculture, since some parts of the country are far better suited to a particular type of agriculture than are others (Barr and Cary, 1992; Williams, 1990). In broadacre cropping (the sector most directly affected by EU and US subsidised exports) three indicators of changing farm financial performance can be calculated from ABARE data for the 33 regions shown in Figure 11.1 in order to demonstrate these very different regional fortunes, namely: changing cash operating surplus; the changing rate of return on labour; and changing debt/equity ratios (Department of Local, 1987; Department of Immigration, 1988).

*Cash operating surplus* is the difference between a farm's total cash costs and total cash receipts. It is a good indicator of the short-term viability of a farm because it excludes non-cash input costs, such as depreciation, and it takes into account delayed payments from previous years, unsold stock and income from sources other than production.

*Rate on return on labour* is the ratio of cash operating surplus to imputed rather than actual labour costs, with family labour notionally being paid at current Federal Pastoral Award rates. If this award rate is interpreted as ensuring that a minimum living wage is paid in pastoral enterprises, then this performance indicator demonstrates the number of times an average broadacre farm earns that living wage for the people employs. Although index values less than 1.0 show regions where, on average, farm operators and their families earning less than a living wage, it can also be argued that index values less than 2.0 identify regions that are far from economically healthy.

*The debt/equity ratio* is a longer-term financial performance measure reflecting the overall solvency of a broadacre farm. It is sometimes referred to as the net worth ratio and measures the financial leverage of a farm, and is a measure of the longer-term financial performance. In a comparative sense, between regions or over time, it measures the improving or deteriorating safety margin that lenders have against possible farm failure.

In order to reveal underlying trends, and to minimise the impact of year-to-year volatility in these three performance indicators, each region's data can be appropriately expressed as regression trends. This procedure generates two summary measures for each performance indicator and for each region; a beginning trend value and an annual rate of change. A third measure, showing current regional differences in performance, can also be generated by

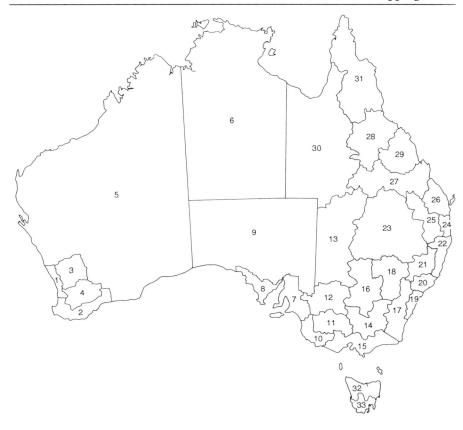

| 1. | Perth | 18. | Dubbo |
| 2. | Bunbury/Albany | 19. | Sydney/Illawarra |
| 3. | Northam | 20. | Hunter |
| 4. | Narrogin | 21. | Tamworth/Armidale |
| 5. | Remote Western Australia | 22. | North Coast NSW |
| 6. | Northern Territory | 23. | Moree |
| 7. | Adelaide | 24. | Brisbane |
| 8. | Whyalla/Port Lincoln | 25. | Toowoomba |
| 9. | Remote South Australia | 26. | Maryborough/Bundaberg/Gladstone |
| 10. | Green Triangle | 27. | Rockingham |
| 11. | Horsham/Bendigo | 28. | Townsville |
| 12. | Riverina/Mildura | 29. | Mackay |
| 13. | Broken Hill | 30. | Mount Isa |
| 14. | Wagga/Albury/Shepparton | 31. | Cairns |
| 15. | Victorian Coastal Region | 32. | Northern Tasmania |
| 16. | Griffith | 33. | Hobart |
| 17. | Canberra | | |

**Figure 11.1**  *Agricultural regions defined by the Office of Local Government and Administrative Services*

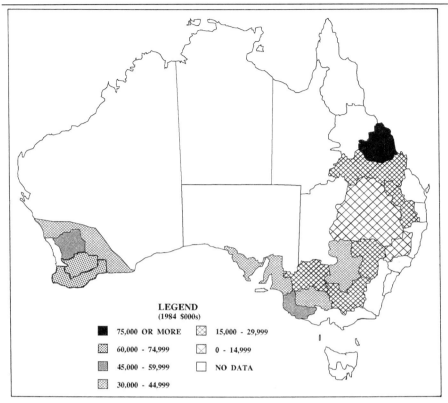

**Figure 11.2**   *Broadacre cropping: 1984–85 to 1990–91. Change in cash operating surplus. Beginning trend value*

averaging the performance data for 1989–90 and 1990–91. The three measures for each financial performance indicator are shown in Figures 11.2 to 11.10.

In the mid-1980s, the indicator of the short-term viability of farms in broadacre cropping — cash operating surplus — showed major regional differentiation from annual levels under A\$15 000 to annual levels over A\$75 000 as shown in Figure 11.2. The ending of the drought in the eastern States saw cash operating surpluses at very low levels in the regions of Tamworth/Armidale, and at quite modest levels (below A\$45 000) in most other wheat-belt regions in Queensland, New South Wales and South Australia. The only exceptions to this pattern were in the Mackay region in Queensland and in the Green Triangle region straddling the Victoria, South Australia, border. In Western Australia the situation was very different. Here farms in the more established wheat-belt regions generated average cash operating surpluses over A\$45 000. Only on the larger properties of the more

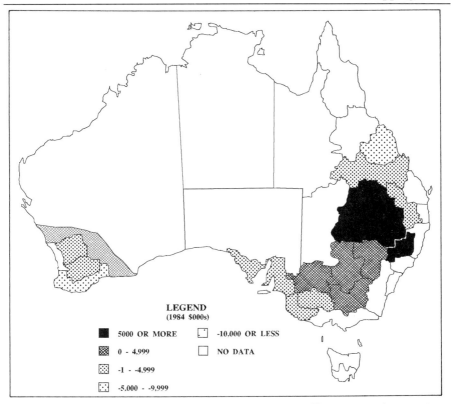

**Figure 11.3**  *Broadacre cropping: 1984–85 to 1990–91. Change in cash operating surplus. Annual rate of change*

recently opened-up, arid, inland periphery of the Western Australian wheat-belt was short-term farm viability lower and at levels comparable with those in eastern States regions.

From this mid-1980s base, the annual rate of change in cash operating surplus has brought about an equalising of regional fortunes as shown in Figure 11.3. The regions of Moree and Tamworth/Armidale have recovered reasonably strongly with average annual increases in cash operating surplus of A$5 000 (in 1984 dollars) having enhanced their short-term viability. Smaller improvements in short-term viability occurred in other parts of New South Wales and also in northern Victoria. However, elsewhere in Victoria, in Queensland and in South Australia, the short-term viability of farms in broadacre cropping deteriorated. The same deterioration occurred in the more established regions of the Western Australian wheat-belt, but not in the regions of the more arid inland fringe in that State.

By the beginning of the 1990s as shown in Figure 11.4, only farms in the

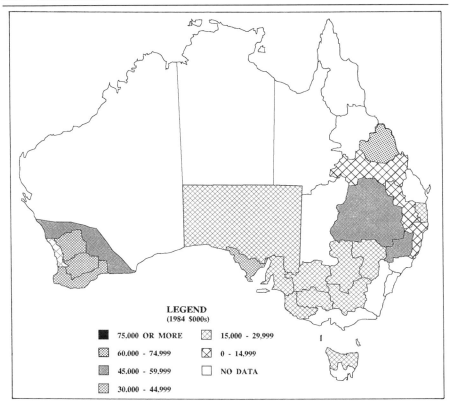

**Figure 11.4**   *Broadacre cropping: 1989–90 and 1990–91. Farm financial performance. Cash operating surplus*

Moree and Tamworth/Armidale regions had significantly improved levels of cash operating surplus. The inland fringe of the Western Australian wheatbelt had also achieved some improvement in this aspect of performance, but only in conjunction with substantial farm amalgamation and farm enlargement (Steward, 1992). In the Mackay region in Queensland, levels of cash operating surplus remained at the relatively high levels of the mid-1980s. However, through all other regions, average cash operating surplus was generally below A$30 000 in the early 1990s, demonstrating widespread weakness in short-term viability in this sector of Australian broadacre production.

As a financial performance indicator rate of return on labour adds a further dimension to the interpretation of regional patterns of short-term viability in broadacre cropping; whether of not farms, on average, are generating a basic living wage for their operators and their families. The information in Figure 11.5 suggests that in the better-performing wheat-belt

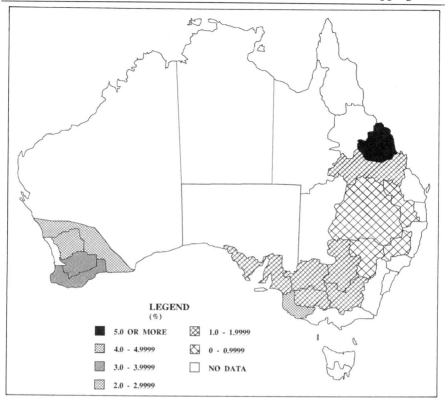

**Figure 11.5**   *Broadacre cropping: 1984–85 to 1990–91. Change in rate of return on labour. Beginning trend value*

regions in the mid-1980s, especially in the Mackay region in Queensland and in all the regions of the Western Australian wheat-belt, the average farm generated an income between four and five times better than a basic living wage. However, in northern New South Wales and southern Queensland (the Moree, Tamworth/Armidale, Dubbo and Toowoomba/Darling Downs regions) the aftermath of the early 1980s drought meant that the average farm did not provide its operators and their families with even a basic living wage (i.e. the rate of return on labour was less than 1.0). If it is argued, however, that an average rate of return on labour greater than 2.0 is necessary to compensate for the problems and costs associated with remote area living (in terms of the greater costs associated with, for example, health, schooling and transport), then only the regions of the western Australian wheat-belt, Mackay and the 'Green Triangle' provided the average farmer with anything better than a basic living wage in the immediately post-drought years of the mid-1980s.

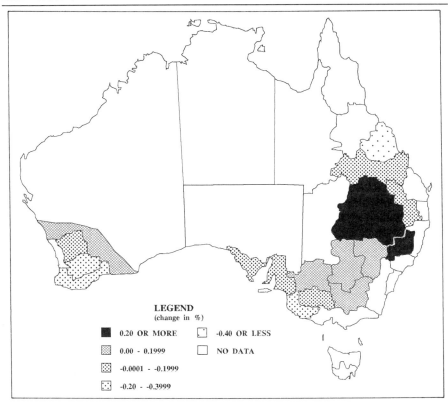

**Figure 11.6**  *Broadacre cropping: 1984–85 to 1990–91. Change in rate of return on labour. Annual rate of change*

From the mid-1980s to the early 1990s and badly affected regions of northern New South Wales and southern Queensland, and to a lesser extent the remaining regions of New South Wales, recovered their income-generating capacity as shown in Figure 11.6. The same recovery also occurred on the inland and more arid margin of the Western Australian wheat-belt. However, elsewhere in Australia, regional rates of return on labour deteriorated. As a result, by the early 1990s no region had an average rate of return on labour greater than 2.0, and for the majority of regions the rate was below 2.0 as shown in Figure 11.7. It would appear, therefore, that through the second half of the 1980s and into the 1990s increasing numbers of farmers in increasing numbers of regions in Australia were being driven towards the poverty line.

The regional picture of changing debt/equity ratios reinforces this pessimistic picture and, owing to the nature of the measure, suggests that the problem is also long-term. In the mid-1980s as shown in Figure 11.8, high

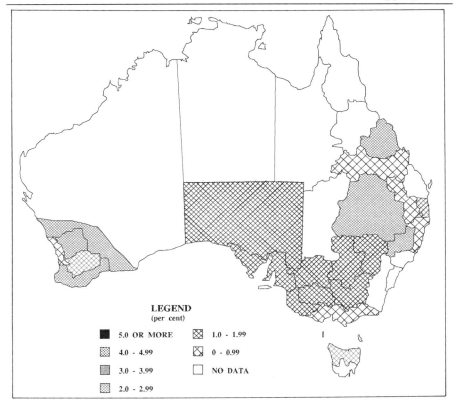

**Figure 11.7**    *Broadacre cropping: 1989–90 and 1990–91. Farm financial performance. Rate of return on labour*

debt/equity ratios of over 24 per cent were recorded in New South Wales and southern Queensland, and again in the regions of Moree, Toowoomba, Dubbo and Griffith. Only in the northern and eastern parts of the Western Australia wheat-belt did debt/equity ratios approach these proportions, but in these cases offset by higher cash operating surpluses as shown in Figure 11.2. In all other regions debt equity ratios in the mid-1980s were below 18 per cent. Between 1984–5 and 1990–91 trends in regional debt levels were downward in most regions of New South Wales and southern Queensland and upwards in all other regions as shown in Figure 11.9. By the early 1990s regional differences in debt/equity ratios had again begun to equalise but at average levels much higher than in the mid-1980s. The ratios were usually over 18 per cent and frequently over 30 per cent as shown by Figure 11.10.

From these regional data four general tendencies are apparent in the financial performance of farms in broadacre cropping in Australia. First, regional differences in indicators of farm financial performance have tended

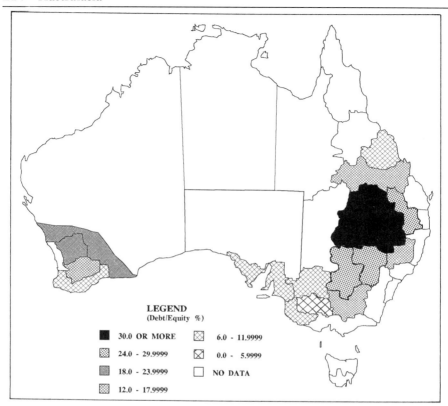

**Figure 11.8**    *Broadacre cropping: 1984–85 to 1990–91. Change in debt/equity ratio. Beginning trend value*

to converge in this period, bringing greater regional uniformity. Second, widespread weakness has emerged in short-term viability as measured by levels of cash operating surplus. Third, rates of return on labour have tended to equalise between regions and have fallen during the period, with the implication that increasing numbers of farm families are being driven towards the poverty line. Fourth, levels of debt, as measured by the debt/equity ratio, have tended to rise and equalise across regions. In some regions they have risen to average levels above 30 per cent, bringing them more closely into line with debt levels in other sectors of the economy.

What has emerged in broadacre cropping in Australia's regions since the mid-1980s is an increasing uniformity of poor financial performance and high debt levels created by depressed international prices resulting from trade conflict beyond the control of Australian farmers. Indeed, it might be suggested that the magnitude of this trade impact has the potential to overwhelm even those regional variations in farm financial performance that

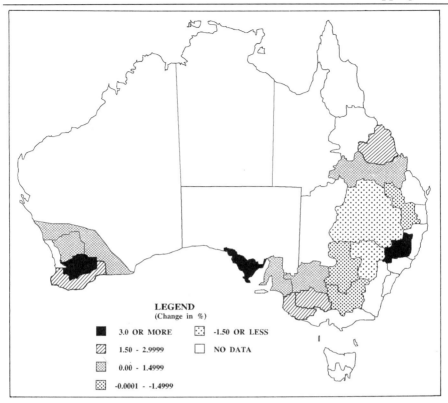

**Figure 11.9**  *Broadacre cropping: 1984–85 to 1990–91. Change in debt/equity ratio. Annual rate of change*

might reasonably have been expected from the different environmental conditions and natural factor endowments that exist between different parts of the country (Parvey, 1988). In the next section of this paper a preliminary attempt will be made, therefore, to measure the impact of environmental conditions on changing levels of farm financial performance in broadacre cropping.

## Explaining regional variations in the financial performance of broadacre cropping

From the preceding analyses of farm financial performance in broadacre cropping it can be suggested that three sets of processes have been responsible for the observed regional patterns and trends:

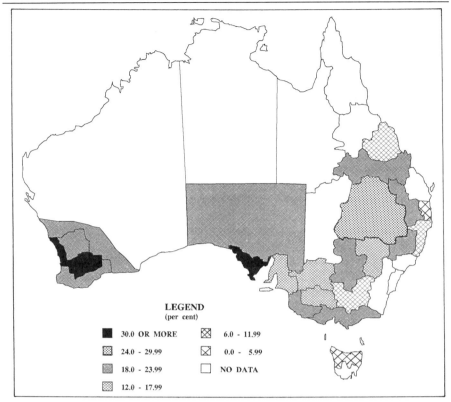

**Figure 11.10**　*Broadacre cropping: 1989–90 and 1990–91. Farm financial performance. Debt/equity ratio*

(1) International market pressures, especially the pressure on prices that has arisen from overseas protectionism and market distortion
(2) Environmental conditions within Australia reflecting, among other things, the differential suitability of regions for broadacre agriculture and the periodic and localised rainfall deficiency and drought that so typifies the continent
(3) The management and operation of family farms and their ability to cope with the fluctuating conditions of both the economic and the physical environments.

Across the regions of Australia, the impact of international market pressures might be expected to be relatively uniform, although farms in some regions might be better placed than those in other regions to cope with the financial problems that are created by these pressures. From the mid-1980s to the early 1990s market conditions can be considered to have exerted considerable downward pressure on prices received by farmers in all regions. These market conditions have also brought a convergence of prices to new and lower levels in all sectors of broadacre farming.

This uniformity contrasts with Australia's very variable physical environment, which significantly affects its capacity to support broadacre farming systems. Williams (1967, p. 30) has identified four environmental factors that shape rural land-use patterns: the availability of soil moisture; topography; the nature of the original vegetation; and soil fertility.

Soil moisture levels affect the length of the growing season and determine where crops can be grown across the continent. Periodic and often localised drought, that so characterises the Australian environment, also influences regional levels and types of broadacre production. In addition, Australia's original vegetation cover has affected land use, with the more sparsely vegetated savannah woodlands having been most easily cleared. Soils and terrain differ markedly from one part of the continent to another, and all too frequently broadacre farming systems have been extended into fragile, marginal environments where they are unsustainable (Barr and Cary, 1992). Pasture improvement through legumes and superphosphate have intensified established farming patterns in some areas and, in conjunction with the use of trace elements, they have also facilitated the extension of the wheat-growing area. This environmental variability might, therefore, be seen as the most important determinant of regional variations in farm financial performance in Australia.

The management and operation of broadacre farms also significantly affects levels of farm financial performance. As 'remnant' organisations that have externalised most of their managerial functions, other than production, to agribusiness, marketing agencies and government (Bowling, 1985; Lawrence, 1987), broadacre family farms have limited capacity to plan for and cope with any forms of external pressure, either environmental or economic. Family lifecycle processes bring periodic financial strains and shifts in the availability of family labour. Intergenerational problems can significantly affect investment decision making, and there are major operational problems that arise from the strongly unequal power relations that exist between family farms and the other enterprises and government agencies and intrumentalities with which they must deal. It can be suggested that while differences in the sizes of properties and the capacity and need to use non-family labour might foster variable levels of financial performance between regions, continuing high levels of family ownership and operation of farms engaged in broadacre agriculture might work in the opposite direction and foster relative uniformity in financial performance between regions.

The data available to test the impact of this range of factors on regional variations in farm financial performance in broadacre cropping are limited, especially data measuring regional differences in farm management, operation and ownership. However, the impact of environmental variables of the type

Table 11.2    *Variables used to identify the environmental characteristics of the agricultural regions drawn up by the Department of Local Government and Administrative Services*

| | Variable | Definition |
|---|---|---|
| (1) | Medium level of water supply | Proportion of the area of a region with mean annual runoff between 5 mm and 50 mm |
| (2) | High level of water supply | Proportion of the area of a region with mean annual runoff greater than 50 mm |
| (3) | Potential for land degradation | Weighted function of gully erosion, dry land salinity and wind erosion for each region on an ordinal scale from 1 (low) to 10 (high) |
| (4) | Plains | Proportion of the area of a region that is plains |
| (5) | Viability for pastoralism | Likelihood of a pastoral enterprise in a region remaining solvent based on experts' judgements. Scale runs from 22 (low) to 75 (high) |
| (6) | Climatic suitability for agriculture | Based on expert judgement of the potential of climatic regions to generate revenue from either cropping or livestock. Scale runs from 1 (low) to 10 (high) |
| (7) | Soil suitability for agriculture | Expert assessment of the capability of soil types to sustain general-purpose cropping assuming a non-limiting climate |
| (8) | Rainfall deficiency | Below-average rainfall affecting more than 50 per cent of a region in any three-month period during a particular year |

*Source*: derived from Parvey (1988)

identified by Williams (1967) can be assessed using data from the CSIRO's Australian Resource Information System (ARIS) (Parvey, 1988), and annual data on rainfall deficiency from the Meteorological Office.

From these sources, eight variables can be constructed for each of the 33 agricultural regions defined by the Office of Local Government (Department of Immigration, 1988). The regional framework as shown in Figure 11.1 is coarse and serves as a limitation to the analysis. The variables, however, measure a number of major environmental characteristics for each region, including available water resources, terrain, soils and climate. They are listed and described in Table 11.2.

These environmental characteristics can be used as independent variables in multiple regression analyses to gain a greater understanding of regional variations in the financial performance of farms engaging in broadacre cropping. For the present study, separate analyses have been undertaken for each indicator for 1984–5, 1990–91 and for the trend rate of change in each performance indicator that had been estimated by simple regression. The results of the analyses are reported in Tables 11.3–11.5. They paint an interesting although somewhat disturbing picture in relation to the sustainability of broadacre agriculture in Australia.

Table 11.3  *Cash operating surplus, environment and broadacre cropping: multiple regression results*

| Variables | Cash operating surplus | | |
|---|---|---|---|
| | | Trend change | |
| | 1984–85 | 1990–91 | 1984–91 |
| Constant | 110342 | 219004 | 25880 |
| $X_1$ % medium water supply | −492 | −42 | −39 |
| $X_2$ % high water supply | +388 | −1 741* | −359 |
| $X_3$ potential for land degradation | +216 | +8 223* | −340 |
| $X_4$ % plains | −456 | −1 720* | −197* |
| $X_5$ pastoral viability | −244 | +97 | +7 |
| $X_6$ climatic suitability | −2 871 | −16 764* | −1 622 |
| $X_7$ % highly productive soil | −445 | −796 | +139* |
| $X_8$ rainfall deficiency | +13 423 | +56 643* | −171 |
| Error term (+/−) | 16 509 | 13 956 | 4 264 |
| $R^2$ | 0.75 | 0.74 | 0.57 |
| $F$ | 3.42** | 3.28 | 1.53 |

* $t$ statistics significant at the 10 per cent level or better
** $F$ statistics significant at the 5 per cent level or better

Table 11.4  *Rate of return on labour, environment and broadacre cropping: multiple regression results*

| Variables | Rate of return on labour | | |
|---|---|---|---|
| | | Trend change | |
| | 1984–5 | 1990–91 | 1984–91 |
| Constant | 5.27 | 5.63 | 2.17 |
| $X_1$ % medium water supply | −0.02 | −0.00 | +0.00 |
| $X_2$ % high water supply | +0.02 | −0.04 | +0.02 |
| $X_3$ potential for land degradation | +0.03 | +0.26 | +0.20 |
| $X_4$ % plains | −0.02 | −0.05* | +0.00 |
| $X_5$ pastoral viability | −0.01 | +0.01 | −0.00 |
| $X_6$ climatic suitability | −0.14 | −0.43* | −0.21 |
| $X_7$ % highly productive soils | −0.02 | −0.02 | −0.02* |
| $X_8$ rainfall deficiency | +0.52 | +1.58* | +0.32* |
| Error term (+/−) | 0.79 | 0.33 | 0.45 |
| $R^2$ | 0.74 | 0.79 | 0.90 |
| $F$ | 3.32 | 4.30** | 10.09** |

* $t$ statistics significant at the 10 per cent level or better
** $F$ statistics significant at the 5 per cent level or better

Table 11.5   *Debt/equity ratio, environment and broadacre cropping: multiple*
*regression results*

|  | Debt/equity ratio | | |
|  | 1984–5 | Trend change 1990–91 | 1984–91 |
| Variables | | | |
|---|---|---|---|
| Constant | 31.23 | 34.43 | −3.29 |
| $X_1$ % medium water supply | −0.12 | −0.08 | +0.03 |
| $X_2$ % high water supply | −0.00 | −0.24 | +0.06 |
| $X_3$ potential for land degradation | −0.11 | 0.92 | +0.06 |
| $X_4$ % plains | −0.21 | −0.08 | +0.04 |
| $X_5$ pastoral viability | +0.11 | −0.01 | −0.01 |
| $X_6$ climatic suitability | −0.04 | +1.09 | +0.51 |
| $X_7$ % highly productive soils | +0.03 | −0.19 | −0.07* |
| $X_8$ rainfall deficiency | −3.59 | +0.40 | −0.22 |
| Error term (+/−) | 5.17 | 6.93 | 11.36 |
| $R^2$ | 0.65 | 0.35 | 0.48 |
| F | 2.08** | 0.61** | 1.04** |

* $t$ statistics significant at the 10 per cent level or better
** no F statistics significant at the 5 per cent level or better

First, only three of the nine equations reported in Tables 11.3–11.5 are
significant at the 5 per cent level. This is, in part, a result of the small number
of regions — only 16 — for which consistent time-series information could be
compiled. However, it also reflects the generally weak relationship between
regional variations in environmental conditions and farm financial performance in broadacre cropping, except in the case of regional rates of return on
labour in the early 1990s.

Second, the weak patterns that emerge from the analyses of cash operating
surpluses and rates of return on labour are almost the opposite of what
would be expected. Beginning with the situation in 1984–5, the regression
results suggest that across Australia's regions the short-term farm financial
performance that is measured by these two indicators was positively
associated with high levels of water supply; as might be expected. However, it
was also positively associated with higher levels of potential land degradation
and the local incidence of drought in that year. What is more, all the other
environmental characteristics were negatively associated with measures of
short-term performance: medium levels of water supply; plains terrain;
pastoral viability; climatic suitability for agriculture; and highly productive
soils. In short, there was a weak tendency (which was strongest for cash
operating surplus) for the best short-term performance to occur in Australia's
more marginal and fragile environments.

Third, by 1990–91 these patterns and relationships had been reinforced, and many of the environmental variables began to show significant partial relationships (at the 10 per cent probability levels) with the cash operating surplus and rates of return on labour indicators. The presence of medium levels of water supply was now negatively associated with performance. In addition, pastoral viability was now positively associated with both measures of short-term performance, possibly reflecting the substitution of wheat for sheep on some properties owing to major falls in wool prices.

The separate analyses for the trend rate change in cash operating surplus and especially rates of return on labour further reinforce this interpretation. However, it is important to stress that the statistical relationships identified here are only weak. Consequently, it is important to emphasise the importance of managerial competency and ability on broadacre farms as a determinant of financial performance quite independent of environmental constraints.

Indeed, there is only one statistically significant, but interpretationally minor, relationship between regional environmental conditions, as they have been measured here, and regional levels and changes in debt/equity ratios in the 1984–5 to 1990–91 period as shown in Table 11.5. In other words, the longer-term solvency of farms engaged in broadacre cropping are largely determined by market pressures and managerial competencies. Environmental factors in the current period of recession are of lesser significance.

## Conclusion

Broadacre agriculture in Australia has always been vulnerable to the fluctuating conditions of international agricultural markets, and that vulnerability has never been stronger than in the years since the mid-1980s. Protectionism in Europe, the USA and Asia has seriously undermined those markets, synchronising downward pressure on prices in all sectors of broadacre farm production.

Since the mid-1980s there has been a widespread and general decline in levels of farm financial performance in Australia in direct response to these external market pressures. In broadacre cropping, regional variations in financial performance have become more uniform as performance in general has declined. Debt levels have risen, and in more and more regions farms in broadacre cropping are now providing their operators and their families with less than a living wage. There is also evidence to suggest that national performance in broadacre cropping has been maintained by shifting production into more marginal regions and environments, although managerial ability on traditionally family owned farms would appear to be an increasingly important determinant of financial performance.

It would appear more urgent than ever to question whether Australia's broadacre farming systems are really viable in the long term — either economically or environmentally. For how long can the Australian broadacre family farm continue as a 'cheap' production system, on the one hand, exploiting the farm family, and, on the other, exploiting fragile and arid environments? It can be contended that in the current economic climate it will be the family farm that will succumb first and, in all probability, will be replaced by corporate operations. If this happens, it will be the environment that will pay the cost as it is exploited to meet short-term financial expediency.

# References

Australian Bureau of Agricultural and Resource Economics, 1988, 'Some major issues affecting the future of Australian agriculture', *Quarterly Review of the Australian Economy*, 10: 38–47

Australian Bureau of Agricultural and Resource Economics, 1991, 'Statistical tables', *Farm Survey Report*, April

Bureau of Agricultural Economics, 1985, *Agricultural Policies in the European Community: Their Origins, Nature and Effects on Production and Trade*, Policy Monograph No. 2, AGPS, Canberra

Backhouse, M., Stoneham, G., Tucker, J. and Walshaw, T., 1989, 'Financial situation in the rural sector', *Farm Surveys Report*, April: 1–14

Barr, N. and Cary, J., 1992, *Greening a Brown Land: The Australian Search for Sustainable Land Use*, Macmillan Education, Melbourne

Bowling, J., 1985, *Technology, Welfare and Intensive Animal Farming: Case Studies of the Poultry and Pig Industries*, Unpublished PhD Thesis, Australian National University, Canberra

Connolly, G., Roper, H. and Barrett, D., 1990, 'Commodity outlook: wool', *Agriculture and Resources Quarterly*, 2: 114–15

Curran, B., Freeman, F. and Sterland, B., 1990, 'Strategic trade policies: implications for Australian industry development', *Agriculture and Resources Quarterly*, 2: 147–56

Department of Immigration, Local Government and Ethnic Affairs, 1988, *A Regionalisation of Australia for Comparative Economic Analysis*, Australian Regional Developments Series, No. 4.6, AGPS, Canberra

Department of Local Government and Administrative Services, 1987, *Regional Variations in Farm Financial Performance*, Australian Regional Developments Series, No. 4.1, AGPS, Canberra

Harris, D. and Dickson, A., 1990, 'Developments in North Asian Beef Markets', paper presented to the North Asia Beef Symposium, National Agricultural and Resources Outlook Conference, Canberra, 30 January–1 February.

Jeans, D.N. (ed.), 1986, *Australia — A Geography*, Vol. 1: *The Natural Environment*, 2nd edition, Sydney University Press, Sydney

Knopke, P., 1990, 'Commodity outlook: farm costs', *Agriculture and Resources Quarterly*, 2: 129

Knopke, P. and Harris, J.M., 1991, 'Changes in input use on Australian farms', *Agriculture and Resources Quarterly*, 3: 230—40

Lawrence, G., 1987, *Capitalism and the Countryside*, Pluto Press, Sydney

Love, G. and Clark, J., 1990, 'The US farm bill: directions for grains', *Agriculture and Resource Quarterly*, 2: 157–65

Miller, G., 1987, *The Political Economy of International Agricultural Policy Reform*, AGPS, Canberra

Parvey, C., 1988, *Economic Aspects of the Physical Environment*, Australian Regional Developments Series, No. 4.4, AGPS, Canberra

Peterson, D.C., Dunne, S.H., Morris, P.C. and Knopke, P., 1991, 'Developments in debt in broadacre agriculture', *Agriculture and Resources Quarterly*, 3: 349–60

Reithmuller, P., Wallace, N. and Tie, G., 1988, 'Government intervention in Japanese agriculture', *Quarterly Review of the Rural Economy*, 10: 155–63

Roberts, I.M., Andrews, N.P. and Hunter, R., 1991, '"Decoupling" and the 1990 US farm bill for grain', *Agriculture and Resources Quarterly*, 3: 203–13

Roberts, I.M., Andrews, N. and Rees, R., 1992, 'Market effects of the 1992 EC reforms of cereals and beef', *Agriculture and Resources Quarterly*, 4: 584–602

Roberts, I.M. and Love, G., 1989, 'Some international effects of the US export enhancement program', *Agriculture and Resources Quarterly*, 1: 170–81

Roberts, I. and Whish-Wilson, P., 1993, 'The US export enhancement program and the Australian wheat industry', *Agriculture and Resources Quarterly*, 5: 228–41

Roper, H.E., Barrett, D.L. and Tran, Q.T., 1990, 'Commodity highlights: wool', *Agriculture and Resources Quarterly*, 2: 247–8

Steward, K., 1992, *The Impact of Restructuring and Recession in the Wheat Belt of Western Australia*, unpublished Honours Dissertation, Department of Geography, University of Western Australia

Taylor, M., 1993, *The Geography of Agricultural Recession: The Financial Performance of Australian Broadacre Farming in the 1980s*, Discussion Paper, Office of Regional Development, Canberra

Tsolakis, D. and Sheales, T., 1990, *Effects of EC Beef and Dairy Policies on Beef Supply, Demand and Trade in the European Community*, ABARE Discussion Paper 90.9, AGPS, Canberra

Tucker, J., Berenger, T. and Backhouse, M., 1990, 'Financial performance of Australian farms', *Farm Surveys Report*, April: 1–12

Webber, M., 1990, 'Garnaut: the implications of northeast Asia for Australian industry', *Australian Journal of International Affairs*, 44: 39–44

Williams, D. (ed.), 1967, *Agriculture in the Australian Economy*, 1st edition, Sydney University Press, Sydney

Williams, D. (ed.), 1990, *Agriculture in the Australian Economy*, 3rd edition, Sydney University Press, Sydney